Lecture Notes in Mathematics

Ilya S. Molchanov

Limit Theorems for Unions of Random Closed Sets

Springer-Verlag
Berlin Heidelberg New York
London Paris Tokyo
Hong Kong Barcelona
Budapest

Author

Ilya S. Molchanov
Department of Mathematics
Kiev Technological Institute of the Food Industry
Vladimirskaya 68
252017 Kiev, Ukraine
and
FB Mathematik
TU Bergakademie Freiberg
Bernhard-v.-Cotta-Str. 2
D-09596 Freiberg, Germany

Mathematics Subject Classification (1991): 60-02, 60D05, 60G55, 60G70, 26E25, 28B20, 52A22

ISBN 3-540-57393-3 Springer-Verlag Berlin Heidelberg New York
ISBN 0-387-57393-3 Springer-Verlag New York Berlin Heidelberg

Printed in Germany

2146/3140-543210 - Printed on acid-free paper

PREFACE

The theory of geometrical probability is, certainly, one of the oldest branches of probability theory. It deals with probability distributions on spaces of geometrical objects (points, lines, planes, triangles, sets etc.) and the corresponding random elements, see Ambartzumian (1990). The notion of a random closed set was introduced by Kendall (1974) and Matheron (1975). Since their studies the concept of probability was defined in a satisfactory manner from the point of view of probability measure on a space of closed sets.

Although a random closed set is a special case of general random elements, random sets have special properties due to the topological structure of the space of closed sets and specific features of set-theoretic operations. Therefore, well-known theorems of classical probability theory gain new meanings and features within the framework of the theory of random sets.

The role and place of limit theorems in probability theory can scarcely be exaggerated. Many important distributions appear as limiting ones with respect to various operations. It is of great interest to derive limit theorems for random sets with respect to set-theoretic operations such as union, intersection or Minkowski (element-wise) addition. It should be noted that limit theorems for random vectors will naturally follow from limit theorems for random sets, since a random vector can be considered to be a single-point random set. On the other hand, limit theorems for random sets gain new features as long as we deal with shapes of limiting random sets and summands.

The limit theorems for random sets have been investigated mostly for the Minkowski addition. The properties of this operation imply that the limiting distribution corresponds to a convex random closed set. Since any convex set can be associated with its support function, limit theorems for Minkowski sums follow from the central limit theorem for sums of random support functions as Banach-space-valued random elements.

In these notes we consider limit theorems for unions of random sets. It should be noted that the union scheme for random sets generalizes the max-scheme for random vectors in a partially-ordered space, whereas Minkowski addition of random sets generalizes the additive scheme for random vectors in a linear space. Limiting random sets for normalized unions of independent identically distributed random sets are naturally said to be union-stable.

It is well-known that the distribution of a random closed set is determined by the corresponding capacity (or hitting) functional on the class of all compacts. This functional is a so-called alternating Choquet capacity of infinite order. Although there are many examples of capacities, sometimes they are not alternating or the corresponding random sets are difficult to construct and simulate. The main stumbling block in the theory of random sets and, especially, in statistics of random sets, is the shortage of convenient models of random sets. In fact, until now only the grain-germ (or Boolean) model provides suitable examples of random sets. In this connection, it should be noted that limit theorems for unions and convex hulls supply us with new models of random sets, which appear as limits.

Unlikely distribution functions of random variables, a principal problem in the theory of random sets is to reduce the number of compacts needed to determine the distribution of a random set by means of its capacity functional on the chosen class. Similar problems are of no interest in classical probability theory, since a distribution function or density are defined naturally on the whole space. The chosen class of compacts then appears in a strong law of large numbers for unions and in definitions of probability metrics for random sets.

Similarly to the max-scheme for random variables or coordinate-wise-maximum-scheme for random vectors, the analysis of unions of random closed sets uses the technique of regularly varying functions. On the other hand, the theory of random sets sparks the theory of regularly varying functions with new concepts such as regularly varying capacities or multivalued regularly varying functions.

The probability metrics method elaborated by Zolotarev (1986) has proved its efficiency in the study of limit theorems for random variables. We define some probability metrics for random closed sets and apply them to limit theorems for unions. The essence of this method lies in proving limit theorems with respect to the most "convenient" metric for the given operation. Then the speed of convergence is estimated with respect to other metrics by the instrumentality of the appropriate inequalities between probability metrics.

Many of the ideas of these notes originate in the pioneering work done by Matheron (1975), who introduced the first notion of union-stability and infinite-divisibility of random sets. Very general notions of infinite divisibility and stability of random sets with respect to various set-theoretic operations were introduced by Trader (1981). Some of the results presented in these notes are closely connected with recent works on general extremal processes, max-stable random vectors and lattice-valued random elements, see Norberg (1986b, 1987), Vervaat (1988), Pancheva (1988), Gerritse (1986, 1990).

The book begins with the introduction of the basic tools and known results on random sets distributions and their weak convergence. Although the book is devoted to the study of limit theorems for unions, in Chapter 2 we present several results on Minkowski sums of random compact sets in the Euclidean space. In Chapter 3 we bring the notions of union-stable and convex-stable random closed sets. Their distributions are characterized in terms of the corresponding capacity or inclusion functionals. In Chapter 4 we prove limit theorems for scaled unions and convex hulls of random sets. Limit theorems for unions of special random sets (random triangles, balls) are considered too. Almost sure stability of unions is investigated in Chapter 5. In Chapter 6 the limit theorems for unions are reformulated in terms of regularly varying multivalued functions, whose definition is introduced too. Chapter 7 is devoted to the development of the probability metrics method in the framework of random sets theory. In the last chapter we discuss several applications. The content of Chapter 8 ranges from the estimates of the volume of random samples and the corresponding statistical tests to the limit theorems for pointwise maxima of random functions and polygonal approximations of convex compact sets.

In each chapter we use notations introduced in it without any comments. While referring to theorems, propositions, examples, formulae etc. from the same chapter we use two-digit notations, e.g., (3.2) designates the second formula from the third section of the same chapter. Otherwise three-digit notations are used, e.g., Theorem 3.1.1 designates Theorem 1.1 from Chapter 3.

I am grateful to Professor V.M.Zolotarev for suggesting the idea of writing these notes and for his further encouragement. These notes appeared as a result of an attempt to generalize the probability metric method for random closed sets. The idea originated in the annual workshop on stability problems for stochastic models organized by V.M.Zolotarev and V.V.Kalashnikov. I thank the organizers and participants of this workshop for helpful comments.

This book was benefited from a lot of discussions with Professor D.Stoyan. His suggestions led to a substantial improvement of the text. The final stage of the work was carried out at the time of my stay at the Technical University Mining Academy of Freiberg. This stay would have been impossible without the financial assistance of the Alexander von Humboldt-Stiftung (Bonn, Germany) and the hospitality of the Mining Academy.

I am indebted to all my colleagues for invitations, comments and discussions of this work at different stages and sending me reprints and preprints, especially, to A.J.Baddeley, N.Cressie, W.F.Eddy, F.Hiai, N.V.Kartashov, V.S.Korolyuk, E.Omey, E.Pancheva, T.Norberg, R.Rebolledo, A.D.Roitgartz, V.Schmidt, F.Streit, W.Vervaat, R.Vitale, W.Weil, M.Zähle and many others.

Special thanks go out to my mother for her invaluable help and constant attention to my research work.

Freiberg, July 7th, 1993 Ilya Molchanov

Contents

Chapter 1

Distributions of Random Closed Sets

1.1 The Space of Closed Sets.

Roughly speaking, a random closed set is a random element in the space of all closed subsets of the basic setting space E. The setting space E in the classical theory of random sets (see Matheron (1975), Stoyan, Kendall and Mecke (1987), Cressie and Laslett (1987) as principal references) is supposed to be locally compact, Hausdorff and separable. It should be noted that Norberg and Vervaat (1989) recently showed that non-Hausdorff E is the natural setting too.

Everywhere below we consider random closed sets in $\mathbb{R}^d$ only, i.e. we suppose E to be equal to $\mathbb{R}^d$. Nevertheless, many results can be easily reformulated for random closed sets in a general finite-dimensional linear space E. The dimension d of the Euclidean space is supposed to be fixed. The Euclidean norm and metric in $\mathbb{R}^d$ are denoted by $\|.\|$ and $\rho(.,.)$ respectively. The ball of radius r centered at x is denoted by $B_r(x)$. We shortly write B_r instead of $B_r(0)$ and B instead of $B_1(0)$.

Define $\mathcal{F}$ to be the family of all closed subsets of $\mathbb{R}^d$ (including the empty set $\emptyset$). Introduce sub-classes of $\mathcal{F}$ by

$$\mathcal{F}^X = \{F \in \mathcal{F} : F \cap X = \emptyset\}, \mathcal{F}_X = \{F \in \mathcal{F} : F \cap X \neq \emptyset\}, \tag{1.1}$$

where $X \subset \mathbb{R}^d$. The class $\mathcal{F}$ is endowed with the topology $\mathbb{T}_f$ (sometimes called *hit-or-miss topology*) generated by

$$\mathcal{F}^K_{G_1,\dots,G_n} = \mathcal{F}^K \cap \mathcal{F}_{G_1} \cap \dots \cap \mathcal{F}_{G_n}, \tag{1.2}$$

where $n \geq 0$, K runs through the class $\mathcal{K}$ of compacts in $\mathbb{R}^d$, $G_1, \dots, G_n$ belong to the family $\mathcal{G}$ of all open sets. It was proven that the space $\mathcal{F}$ furnished with the hit-or-miss topology is compact, separable and Hausdorff, see Matheron (1975).

A sequence of closed sets $F_n, n \geq 1$, converges in $\mathbb{T}_f$ to a certain closed set F if and only if the following conditions are valid

(F1) if $K \cap F = \emptyset$ for a certain compact K, then $K \cap F_n = \emptyset$ for all sufficiently large n;

(F2) if $G \cap F \neq \emptyset$ for a certain open set G, then $G \cap F_n \neq \emptyset$ for all sufficiently large n.

We then write $F = \mathcal{F}-\lim F_n$ or $F_n \xrightarrow{\mathcal{F}} F$.

Let $\mathbb{T}_k$ be the topology on $\mathcal{K}$ induced by $\mathbb{T}_f$. To ensure the convergence of a sequence K_n, $n \geq 1$, of compact sets in $\mathbb{T}_k$ an additional condition is required:

(F3) there exists a compact K' such that $K_n \subseteq K'$ for all $n \geq 1$.

We denote $K = \mathcal{K}-\lim K_n$ in case K_n converges to K in $\mathbb{T}_k$.

The convergence of compact sets in $\mathbb{T}_k$ can be metrized by means of the *Hausdorff metric* ρ_H on $\mathcal{K}$. The Hausdorff distance between two compacts K and K_1 is defined as

$$\rho_H(K, K_1) = \inf\{\varepsilon > 0 : K \subseteq K_1^\varepsilon, K_1 \subseteq K^\varepsilon\}, \tag{1.3}$$

where

$$K^\varepsilon = \cup\{B_\varepsilon(x) : x \in K\} = K \oplus B_\varepsilon(0)$$

is the ε-envelope of K, $\oplus$ is the Minkowski addition (see Section 1.5). The Hausdorff distance between two closed sets is defined similarly. However, it can be infinite.

The upper limit $\mathcal{F}-\limsup F_n$ is the largest closed set F which satisfies the condition **(F1)**. Similarly, $\mathcal{K}-\limsup$ is defined by combining **(F1)** and **(F3)**.

Lemma 1.1 *Let K_n, $n \geq 1$, be a sequence of compact sets. Then $K \subseteq \mathcal{K}-\limsup K_n$ if and only if*

$$\varepsilon_n = \inf\{\varepsilon > 0\colon K \subseteq K_n^\varepsilon\} \to 0 \quad as \quad n \to \infty.$$

PROOF. Let $\varepsilon_n \to 0$ as $n \to \infty$. For any x from K there exists a sequence of points $x_n \in K_n$, $n \geq 1$, such that $\|x - x_n\| \leq \varepsilon_n$. Thus, $x_n \to x$ as $n \to \infty$, so that $x \in \mathcal{K}-\limsup K_n$.

Let $K \subseteq \mathcal{K}-\limsup K_n$. Suppose that $\varepsilon_n \geq \delta > 0$, $n \geq n_0$. Then there exist points $x_n \in K$, $n \geq n_0$, such that $B_\delta(x_n) \cap K_n = \emptyset$. Without loss of generality suppose that $x_n \to x_0 \in K$ as $n \to \infty$. Then $B_{\delta/2}(x_0) \cap K_n = \emptyset$, $n \geq n_0$, i.e. $x_0 \notin \mathcal{K}-\limsup K_n$. Hence $\varepsilon_n \to 0$ as $n \to \infty$. □

For later use we denote by $\bar{M}$, Int$M$, ∂M, M^c, conv(M) respectively the closure, interior, boundary, complement in $\mathbb{R}^d$ and the convex hull of any set $M \subset \mathbb{R}^d$.

A set M is said to be *canonically closed* if M coincides with the closure of its interior, i.e. $M = \overline{\text{Int}M}$.

1.2 Random Closed Sets and Capacity Functionals.

According to what has been said, a *random closed set* is an $\mathcal{F}$-valued random element. To complete this definition the class $\mathcal{F}$ is endowed with the Borel σ-algebra σ_f generated by $\mathbb{T}_f$. Then a random element in $(\mathcal{F}, \sigma_f)$ is said to be a random closed set (RACS). Here are several examples of random closed sets: random points and point processes, random spheres and balls, random half-spaces and hyperplanes etc.

The distribution of a random closed set A is described by the corresponding probability measure $\mathbf{P}$ on σ_f. In this connection

$$\mathbf{P}\left\{\mathcal{F}^K \cap \mathcal{F}_{G_1} \cap \ldots \cap \mathcal{F}_{G_n}\right\} = \mathbf{P}\left\{A \cap K = \emptyset, A \cap G_1 \neq \emptyset, \ldots, A \cap G_n \neq \emptyset\right\}.$$

Clearly, these probabilities determine the measure $\mathbf{P}$ on σ_f. Fortunately, $\mathbf{P}$ is determined also by its values on $\mathcal{F}_K$ for K running through $\mathcal{K}$ only. Let $T(K)$ be equal to $\mathbf{P}(\mathcal{F}_K)$, i.e.

$$T(K) = \mathbf{P}\{A \cap K \neq \emptyset\}, K \in \mathcal{K}. \tag{2.1}$$

The functional T is said to be the *capacity* (or hitting) *functional* of A. Sometimes we write $T_A(K)$ instead of $T(K)$. Considered as a function on $\mathcal{K}$ the capacity functional T is an *alternating Choquet capacity* of infinite order (briefly Choquet capacity). Namely, T has the following properties:

(T1) T is upper semi-continuous on $\mathcal{K}$, i.e. $T(K_n) \downarrow T(K)$ in case $K_n \downarrow K$ as $n \to \infty$.

(T2) The following functionals recurrently defined by

$$\begin{aligned} S_1(K_0; K) &= T(K_0 \cup K) - T(K_0) \\ \dots & \quad \dots \\ S_n(K_0; K_1, ..., K_n) &= S_{n-1}(K_0; K_1, ..., K_{n-1}) - S_{n-1}(K_0 \cup K_n; K_1, ..., K_{n-1}) \end{aligned}$$

are non-negative for all $n \geq 0$ and $K_0, K_1, ..., K_n$ from $\mathcal{K}$.

The value of $S_n(K_0; K_1, ..., K_n)$ is equal to the probability that A misses K_0 but hits $K_1, ..., K_n$. In particular, T is increasing, since S_1 is non-negative.

The properties of T resemble those of the distribution function. Property **(T1)** is the same as the right-continuity and **(T2)** is the extension of the notion of monotonicity. However, in contrast to measures, the functional T is not additive, but only subadditive.

EXAMPLE **2.1** Let $A = (-\infty, \xi]$ be a random set in $\mathbb{R}^1$, where ξ is a random variable. Then $T(K) = \mathbf{P}\{\xi > \inf K\}$ for all $K \in \mathcal{K}$.

EXAMPLE **2.2** Let $A = \{\xi\}$ be a single-point random set in $\mathbb{R}^d$. Then $T(K)$ is equal to $\mathbf{P}\{\xi \in K\}$ and coincides with the corresponding probability distribution of ξ. It can be proven that the capacity functional T is additive iff A is a single-point random set.

The powerful result derived by Matheron (1975) and Kendall (1974) establishes one-to-one correspondence between Choquet capacities and distributions of random closed sets.

Theorem 2.3 (Choquet) *Let T be a functional on $\mathcal{K}$. Then there is a (necessary unique) distribution* $\mathbf{P}$ *on $\mathcal{F}$ with*

$$\mathbf{P}\{\mathcal{F}_K\} = T(K), \quad K \in \mathcal{K},$$

if and only if T is an alternating Choquet capacity of infinite order such that $0 \leq T(K) \leq 1$ *and* $T(\emptyset) = 0$.

Capacity functionals play in the theory of random sets the same role as distribution functions in classical probability theory. However, the class $\mathcal{K}$ of all compacts is too large to define efficiently the capacity functional on it. In this connection an important problem arises to reduce the class of test sets needed. That is to say, is the distribution of a random closed set determined by the values $T(K)$, $K \in \mathcal{M}$, for a certain class $\mathcal{M} \subset \mathcal{K}$?

It was proven in Molchanov (1983) that if realizations of a random set belong to a certain sub-class $\mathfrak{S} \subset \mathcal{F}$ then this extra knowledge reduces the class $\mathcal{M}$ of test sets needed.

Theorem 2.4 *Let $\mathfrak{S} \subset \mathcal{F}$, and let $\mathcal{M} \subset \mathcal{K}$. Suppose that the following conditions are valid.*

1. *$\mathcal{M}$ is closed with respect to finite unions.*

2. *There exists a countable sub-class $\mathfrak{B} \subset \mathcal{G}$ such that any compact K from $\mathcal{M}$ is the limit of a decreasing sequence of sets from $\mathfrak{B}$, and also any G from $\mathfrak{B}$ is the limit of an increasing sequence from $\mathcal{M}$.*

3. *For any $G \in \mathfrak{B} \cup \{\emptyset\}$, $K_1, ..., K_n \in \mathcal{M}$, $n \geq 0$, the class*

$$\mathcal{F}^G_{K_1,...,K_n} \cap \mathfrak{S}$$

is non-empty, provided $K_i \setminus G$ is non-empty for all $1 \leq i \leq n$.

4. *The σ-algebra σ_m generated by*

$$\{\mathcal{F}^K_{G_1,...,G_n} \cap \mathfrak{S}\colon K \in \mathcal{M} \cup \{\emptyset\}, G_i \in \mathfrak{B}, 1 \leq i \leq n\}$$

coincides with the σ-algebra $\sigma_f \cap \mathfrak{S} = \{\mathcal{A} \cap \mathfrak{S}\colon \mathcal{A} \in \sigma_f\}$ induced by σ_f on the class $\mathfrak{S}$.

Let $\bar{\mathfrak{S}}$ be the closure of $\mathfrak{S}$ in $\mathbb{T}_f$. Then the functional T on $\mathcal{M}$ is a Choquet capacity of infinite order on $\mathcal{M}$ (i.e. the conditions **(T1)-(T2)** *are valid on $\mathcal{M} \cup \{\emptyset\}$) such that $0 \leq T \leq 1$ and $T(\emptyset) = 0$ if and only if there is a (necessary unique) probability* **P** *on σ_m such that*

$$\mathbf{P}\{\mathcal{F}_K \cap \bar{\mathfrak{S}}\} = T(K), K \in \mathcal{M}.$$

In general, the distribution of any random closed set is determined by the values of its capacity functional on the class $\mathcal{K}_{ub}$ of all finite unions of balls of positive radii, or on the class $\mathcal{K}_{up}$ of all finite unions of parallelepipeds, see Salinetti and Wets (1986), Lyashenko (1983). Norberg (1989) established deep relations between topological properties of continuous partially ordered sets and distributions of random closed sets.

The capacity functional T is said to be *maxitive* if

$$T(K_1 \cup K_2) = \max(T(K_1), T(K_2))$$

for all compacts K_1, K_2. Such capacities arise naturally in the theory of extremal processes, see Norberg (1986b, 1987).

EXAMPLE **2.5** Define a maxitive capacity T by

$$T(K) = \sup\{f(x)\colon x \in K\},$$

where $f: \mathbb{R}^d \to [0,1]$ is an upper semi-continuous function. Then T describes the distribution of the random set A defined as

$$A = \{x \in \mathbb{R}^d : f(x) \geq \eta\},$$

where η is a random variable uniformly distributed on $[0,1]$.

A random closed set A is said to be *stationary* if A and $A+x$ coincide in distribution, whatever x in $\mathbb{R}^d$ may be. Similarly, A is *isotropic* if A has the same distribution as its any non-random rotation. Of course, the capacity functional of a stationary (isotropic) random set is shift-invariant (rotation-invariant).

A random set is said to be *compact* if its realizations are almost surely compact.

1.3 Convex Random Sets.

Define $\mathcal{C}$ to be the class of convex closed sets in $\mathbb{R}^d$, and let $\mathcal{C}_0 = \mathcal{C} \cap \mathcal{K}$ be the class of all convex compact sets. A random closed set is said to be *convex* if its realizations are almost surely convex, i.e. A belongs to $\mathcal{C}$ almost surely. Of course, the distribution of any convex random closed set A is determined by the corresponding capacity functional (2.1). Fortunately, the additional properties of the realizations of A (see Theorem 2.4) yield the reduction of the class of test compacts needed. The following result is due to Vitale (1983). It was proven independently by Molchanov (1983), see also Trader (1981).

Theorem 3.1 *The distribution of any convex compact random set A is determined uniquely by the values of the functional*

$$t(K) = \mathbf{P}\{A \subset K\}$$

for K running through the class $\mathcal{C}_0$ of convex compact sets.

PROOF. Check the conditions of Theorem 2.4. Having considered a single-point compactification $E' = \mathbb{R}^d \cup \{\omega\}$, we can regard A to be a convex RACS in the compact space E'. Since A is supposed to be compact, it misses $\{\omega\}$ almost surely. Let $\mathcal{M}$ be the class of complements to all open bounded convex sets in $\mathbb{R}^d$, and let $\mathfrak{B}$ be the class of complements to convex polyhedrons with rational vertices. It is easy to show that the first and the second conditions of Theorem 2.4 are valid. The third one is valid too, since for all $G \in \mathfrak{B} \cup \{\emptyset\}$, $K_1, ..., K_n \in \mathcal{M}$ and x_i belonging to $K_i \setminus G$, $1 \leq i \leq n$, the convex hull of $\{x_1, ..., x_n\}$ misses G, so that

$$\mathcal{F}^G_{K_1,...,K_n} \cap \mathcal{C}_0 \neq \emptyset.$$

Verify the fourth condition. Let K be a compact set, and let $F \in \mathcal{F}^K \cap \mathcal{C}_0$. Then $F \in \mathcal{F}^{K_1} \cap \mathcal{C}_0$ for a certain K_1 from $\mathcal{M}$. E.g., K can be chosen to be the complement to a certain bounded neighborhood $U(F)$ of F such that $U(F) \cap K = \emptyset$.

Let $F \in \mathcal{F}_G \cap \mathcal{C}_0$ for a certain open G, and let

$$x_0 = (x_{01}, ..., x_{0d}) \in F \cap G.$$

Pick $\delta > 0$ such that

$$G_0 = \left\{ x = (x_1, \ldots, x_d) \colon \max_{1 \le i \le d} |x_i - x_{0i}| < \delta \right\} \subseteq G.$$

For each collection of numbers $l_i = \pm 1$, $1 \le i \le d$, define

$$\begin{aligned} H_j^\varepsilon &= H_{l_1,..,l_d}^\varepsilon \\ &= \left\{ x = (x_1, \ldots, x_d) \colon \sum_{i=1}^{d} (x_i - x_{0i}) l_i > 1 - \varepsilon \right\}, \ \varepsilon > 0, \ 1 \le j \le 2^d. \end{aligned}$$

If ε is sufficiently small, then every convex set, which misses G_0^c and hits H_j^ε, $1 \le j \le 2^d$, also contains x_0. Observe that G_0^c belongs to $\mathcal{M}$. Thus

$$F \in \mathcal{F}^{G_0^c}_{H_1^\varepsilon, \ldots, H_{2^d}^\varepsilon} \cap \mathcal{C}_0 \subseteq \mathcal{F}_G \cap \mathcal{C}_0,$$

whence $\sigma_m = \sigma_f \cap \mathcal{C}_0$.

By Theorem 2.4, there exists the unique probability measure $\mathbf{P}$ on σ_f such that $\mathbf{P}\left\{\mathcal{F}_K \cap \bar{\mathcal{C}}_0\right\} = T(K)$, $K \in \mathcal{M}$. The closure $\bar{\mathcal{C}}_0$ consists of also convex sets containing the point $\{\omega\}$ (i.e. $\bar{\mathcal{C}}_0 = \mathcal{C}$). However, since the random set A is compact, the corresponding probability $\mathbf{P}$ is concentrated within $\mathcal{C}_0$. Thus, $\mathbf{P}\{\mathcal{F}_K \cap \mathcal{C}_0\} = T(K)$ for each compact K. Then the distribution of A is determined by the values $\mathbf{P}\{A \subseteq K^c\}$, whence the statement of Theorem easy follows. □

The functional $\mathfrak{t}(K)$, $K \in \mathcal{C}_0$, is naturally extended onto the class $\mathcal{C}$ by

$$\mathfrak{t}(F) = \mathbf{P}\{A \subset F\}, F \in \mathcal{C}. \tag{3.1}$$

This functional $\mathfrak{t}$ is said to be the *inclusion functional* of A. It is a so-called monotone capacity of infinite order (see Choquet, 1953/54). In other words, it satisfies the following conditions.

(I1) $\mathfrak{t}$ is upper semicontinuous, i.e. $\mathfrak{t}(F_n) \to \mathfrak{t}(F)$ if $F_n \downarrow F$ as $n \to \infty$ for F_n, F belonging to $\mathcal{C}$, $n \ge 1$.

(I2) The recurrently defined functionals

$$\begin{aligned} S_1^{\mathfrak{t}}(F; F_1) &= \mathfrak{t}(F) - \mathfrak{t}(F \cap F_1) \\ \cdots & \quad \cdots \\ S_n^{\mathfrak{t}}(F; F_1, \ldots, F_n) &= S_{n-1}^{\mathfrak{t}}(F; F_1, \ldots, F_{n-1}) - S_{n-1}^{\mathfrak{t}}(F \cap F_n; F_1, \ldots, F_{n-1}) \end{aligned}$$

are non-negative, whatever $n \ge 1$ and $F, F_1, \ldots, F_n$ from $\mathcal{C}$ may be.

In fact, $S_n^{\mathfrak{t}}(F; F_1, \ldots, F_n)$ is the probability that $A \subseteq F$ and $A \not\subseteq F_i$, $1 \le i \le n$. Note that $\mathfrak{t}$ is expressed in terms of the capacity functional T by means of

$$\mathfrak{t}(F) = \mathbf{P}\{A \subseteq F\} = T(F^c), F \in \mathcal{C}. \tag{3.2}$$

The following example shows that the distribution of a non-compact convex RACS, in general, cannot be determined via the functional $\mathfrak{t}$ on $\mathcal{C}$.

EXAMPLE **3.2** Let A be the half-space which touches the unit ball B_1 at a random point uniformly distributed on its boundary. Then $\mathfrak{t}(F) = 0$ whenever $F \in \mathcal{C}$, $F \neq \mathbb{R}^d$. Thus, the inclusion functional of A coincides with the inclusion functional of the set $A = \mathbb{R}^d$.

Nevertheless, we sometimes consider the functional $\mathfrak{t}(F)$, $F \in \mathcal{C}$, even for unbounded A. If A is non-convex, then this functional does not determine the distribution of A, but $\mathrm{conv}(A)$.

For any convex F define the *support function*

$$s_F(u) = \sup\{u \cdot v\colon\ v \in F\}, \tag{3.3}$$

where $u \cdot v$ is the scalar multiplication, u runs through the unit sphere $\mathbb{S}^{d-1}$ in $\mathbb{R}^d$. The function s_F is allowed to take infinite values if F is unbounded. Of course, s_F is finite everywhere iff F is compact.

If A is a convex compact random set, then $s_A(u)$ is the random element in the space $\mathbf{C}(\mathbb{S}^{d-1})$ of continuous functions on $\mathbb{S}^{d-1}$.

Let $\mathcal{H}$ be the class of all finite intersections of half-spaces in $\mathbb{R}^d$.

Proposition 3.3 *The distribution of a compact convex random set is determined by the values of its inclusion functional on* $\mathcal{H}$.

PROOF. The statement follows from the fact that the values $\mathfrak{t}(F)$ for F running through $\mathcal{H}$ determine the finite-dimensional distributions of the random process $s_A(u)$, $u \in \mathbb{S}^{d-1}$. □

1.4 Weak Convergence of Random Closed Sets.

Weak convergence of random sets is a particular case of weak convergence of probability measures, since a random closed set is associated with a certain probability measure on σ_f. A sequence of random closed sets A_n, $n \geq 1$, is said to *converge weakly* if the corresponding probability measures $\mathbf{P}_n$, $n \geq 1$, converge weakly in the usual sense, see Billingsley (1968). Namely,

$$\mathbf{P}_n(\mathfrak{A}) \to \mathbf{P}(\mathfrak{A}) \ \text{ as } \ n \to \infty \tag{4.1}$$

for each $\mathfrak{A} \in \sigma_f$ such that $\mathbf{P}(\partial\mathfrak{A}) = 0$ for the boundary of $\mathfrak{A}$ with respect to $\mathbb{T}_f$ (i.e. $\mathfrak{A}$ is a continuity set for the limiting measure).

However, it is rather difficult to check (4.1) for all $\mathfrak{A}$ from σ_f. The first natural reduction is in letting $\mathfrak{A}$ to be equal to $\mathcal{F}_K$ for K running through $\mathcal{K}$. It was proven in Lyashenko (1983) and Salinetti and Wets (1986) that the class $\mathcal{F}_K$ is a continuity set for $\mathbf{P}$ if

$$\mathbf{P}\left\{\mathcal{F}_K\right\} = \mathbf{P}\left\{\mathcal{F}_{\mathrm{Int}K}\right\}.$$

In other words,

$$\mathbf{P}\left\{A \cap K \neq \emptyset, A \cap \mathrm{Int}K = \emptyset\right\} = 0 \tag{4.2}$$

for the corresponding limiting random closed set A. In terms of the limiting capacity functional we get

$$T_A(K) = T_A(\mathrm{Int}K), \tag{4.3}$$

where

$$T_A(\mathrm{Int}K) = \sup\{T(K'):\ K' \in \mathcal{K}, K' \subset \mathrm{Int}K\}.$$

The class of compacts satisfying (4.2) or (4.3) is denoted by $\mathcal{S}_T$, i.e.

$$\mathcal{S}_T = \{K \in \mathcal{K}:\ T_A(K) = T_A(\mathrm{Int}K)\}. \tag{4.4}$$

Then A_n converges weakly to A if

$$T_{A_n}(K) \to T_A(K) \ \text{ as } \ n \to \infty$$

for each K belonging to $\mathcal{S}_T$. Thus, the pointwise convergence of capacity functionals on $\mathcal{S}_T$ implies the weak convergence of the corresponding probability measures on σ_f.

Further reduction is due to Salinetti and Wets (1986), Norberg (1984). The class $\mathcal{S}_T$ can be replaced with the class $\mathcal{K}_{ub} \cap \mathcal{S}_T$, where $\mathcal{K}_{ub}$ is the class of finite unions of balls having positive radii. In turn, $\mathcal{K}_{ub}$ can be reduced to the countable class $\mathcal{K}_{ub\mathbb{Q}}$ of finite unions of balls with rational midpoints and positive rational radii. Recent results on the convergence of random sets and related topics from the theory of semi-lattices are discussed in Norberg (1989).

In general, the class $\mathcal{M} \subseteq \mathcal{K}$ is said to *determine the weak convergence* if the pointwise convergence of capacity functionals on $\mathcal{M} \cap \mathcal{S}_T$ yields the weak convergence of distributions of random closed sets.

It was proven in Lyashenko (1983) that the class $\mathcal{K}_{up}$ of finite unions of parallelepipeds and even the class $\mathcal{K}_{up\mathbb{Q}}$ of unions of parallelepipeds with rational vertices also determine the weak convergence of random closed sets.

For general random sets there is likely no possibilities of further essential shortening of the class determining the weak convergence. Nevertheless, for compact convex random sets a further reduction is possible. Namely, the weak convergence of compact convex random sets is characterized by the pointwise convergence of capacity (or inclusion) functionals on a smaller class.

Consider convex compact random sets Z_n, $n \geq 1$, $\tilde{Z}$ with the inclusion functionals $\mathfrak{t}_n$, $n \geq 1$, $\tilde{\mathfrak{t}}$ and the corresponding probability measures $\mathbf{P}_n$, $\mathbf{P}$ on $\mathcal{C}_0$ furnished with the σ-algebra induced by σ_f. We say that Z_n converges weakly to $\tilde{Z}$ if $\mathbf{P}$ converges weakly to $\mathbf{P}$ in the usual sense. Since the class $\mathcal{C}_0$ is measurable, this fact yields the weak convergence of the random sets distributions on $(\mathcal{F}, \sigma_f)$.

The following theorem shows that the pointwise convergence of inclusion functionals on $\mathcal{C}_0$ implies the weak convergence of random compact convex sets, see Molchanov (1993d).

Theorem 4.1 *The convex compact random set Z_n converges weakly to the random closed set $\tilde{Z}$ if, for any K from $\mathcal{C}_0$,*

$$\mathfrak{t}_n(K) \to \tilde{\mathfrak{t}}(K) \ \text{ as } \ n \to \infty, \tag{4.5}$$

where $\mathfrak{t}_n, \tilde{\mathfrak{t}}$ are the inclusion functionals of random sets Z_n and $\tilde{Z}$ respectively.

We begin with a lemma.

Lemma 4.2 *Let $B_r(x_0) \subset \mathrm{Int} B_R(0)$. Then there exist sets $F_1, \ldots, F_n$ belonging to $\mathcal{C}_0$ such that $M \cap B_r(x_0) \neq \emptyset$ for any $M \in \mathcal{C}$, $M \subset B_R(0)$, provided $M \not\subset F_i$, $1 \leq i \leq n$.*

PROOF. Pick points $u_1, \ldots, u_m$ from the unit sphere $\mathbb{S}^{d-1}$ such that for any $(d-1)$-dimensional plane ℓ crossing x_0

$$\ell \cap B_R(0) \subset H_r^+(u_i) \cap H_r^-(u_i) \tag{4.6}$$

for some u_i, where

$$\begin{aligned} H_r^+(u) &= \{x \in B_R(0) \colon (x - x_0) \cdot u \leq r\}, \\ H_r^-(u) &= \{x \in B_R(0) \colon (x - x_0) \cdot u \geq -r\}. \end{aligned}$$

Denote $F_{2i} = H_r^+(u)$, $F_{2i-1} = H_r^-(u)$ for $1 \leq i \leq m$, $n = 2m$. If $M \not\subset F_i$, $1 \leq i \leq n$, then there are points

$$x_i \in M \setminus H_r^+(u_i),\ y_i \in M \setminus H_r^-(u_i),\ 1 \leq i \leq m,$$

and also

$$M \supset M_1 = \mathrm{conv}\,\{x_1, \ldots, x_m, y_1, \ldots, y_m\}.$$

Suppose that $x_0 \notin M_1$. Then $M_1 \subset H$ for a certain $(d-1)$-dimensional hyperplane ℓ hitting x_0 and dividing $\mathbb{R}^d$ into half-spaces H and H'. However, (4.6) is valid for a certain i, whence the points x_i, y_i lie in the different half-spaces H and H', i.e. $M_1 \not\subset H$. Therefore, $x_0 \in M$, that is M hits $B_r(x_0)$. □

PROOF OF THEOREM 4.1. Denote for any $X \subset \mathbb{R}^d$

$$\mathcal{C}_X = \{M \in \mathcal{C}_0 \colon M \subset X\},\ \mathcal{C}^X = \{M \in \mathcal{C}_0 \colon M \not\subset X\}. \tag{4.7}$$

The family

$$\mathcal{U} = \{\mathcal{C}_F \cap \mathcal{C}^{F_1} \cap \cdots \cap \mathcal{C}^{F_n} \colon\ F, F_i \in \mathcal{C}, 1 \leq i \leq n, n \geq 0\}$$

is closed with respect to intersections. It follows from (4.5) and **(I2)** that for any $\mathfrak{A}$ from the family $\mathcal{U}$

$$\mathbf{P}_n(\mathfrak{A}) = S_n^{\mathfrak{t}_n}(F; F_1, \ldots, F_n) \to \mathbf{P}(\mathfrak{A}) = S_n^{\mathfrak{t}}(F; F_1, \ldots, F_n) \text{ as } n \to \infty. \tag{4.8}$$

Let K be a compact set, and let $M \in \mathcal{F}^K \cap \mathcal{C}_0$. Then $M \subset \mathrm{Int} F$ and $F \cap K = \emptyset$ for some convex compact F. Hence

$$M \in \mathcal{C}_{\mathrm{Int} F} \subset \mathcal{C}_F \subset \mathcal{F}^K.$$

Let G be any open set such that $M \in \mathcal{F}_G \cap \mathcal{C}_0$, i.e. M hits G. If $M \cap G$ consists of only one point x_0, then $M = \{x_0\}$ and

$$M \in \mathcal{C}_{\mathrm{Int} B_r(x_0)} \subset \mathcal{C}_{B_r(x_0)} \subset \mathcal{F}_G$$

for a certain ball $B_r(x_0) \subset G$.

Suppose that the set $M \cap G$ consists of at most $m < d$ points $x_1, ..., x_m$ such that the vectors $x_j - x_1$, $2 \leq j \leq m$, are linearly independent. Then M is a subset of a certain hyperplane ℓ, containing these points. It is evident that $M \subset B_R(0)$, $B_{2r}(x_0) \subset G$, $B_{2r}(x_0) \cap \ell \subset M$ for some $r, R > 0$. Having applied Lemma 4.2 to the subspace ℓ and balls $B_R(0) \cap \ell$ and $B_r(x_0) \cap \ell$, we obtain the corresponding sets $F_1, ..., F_n$. Put

$$\tilde{F}_i = \{x \in B_R(0)\colon \mathrm{pr}_\ell x \in F_i\},$$

where $\mathrm{pr}_\ell x$ is the projection of x on ℓ. Let F be the intersection of $B_R(0)$ and the ε-envelope of ℓ. Then

$$M \in \mathcal{C}_{\mathrm{Int}F} \cap \mathcal{C}^{\tilde{F}_1} \cap \cdots \cap \mathcal{C}^{\tilde{F}_n} \subset \mathcal{C}_F \cap \mathcal{C}^{\tilde{F}_1} \cap \cdots \cap \mathcal{C}^{\tilde{F}_n} \subset \mathcal{F}_G.$$

If $M \cap G$ consists of the ball $B_{2r}(x_0)$ for some $r > 0$, then Lemma 4.2 immediately yields

$$M \in \mathcal{C}^{F_1} \cap \cdots \cap \mathcal{C}^{F_n} \subset \mathcal{F}_G$$

for $F_1, ..., F_n$ belonging to $\mathcal{C}_0$.

As stated in Section 1.1, the classes

$$\mathcal{A} = \mathcal{F}^K \cap \mathcal{F}_{G_1} \cap \cdots \cap \mathcal{F}_{G_n}$$

for K running through $\mathcal{K}$ and all open sets $G_1, ..., G_n$, $n \geq 0$, form the base of the topology $\mathbb{T}_f$ and generate the Borel σ-algebra σ_f on $\mathcal{F}$. Observe that $\mathcal{C}^F$, $\mathcal{C}_{\mathrm{Int}F}$ are open in this topology for each F from $\mathcal{C}_0$. We have proved above that for such a class $\mathcal{A}$ and any M belonging to $\mathcal{A} \cap \mathcal{C}_0$

$$M \in \mathfrak{A}_0 \subset \mathfrak{A} \subset \mathcal{A} \cap \mathcal{C}_0$$

for a certain $\mathfrak{A} \in \mathcal{U}$, where $\mathfrak{A}_0$ is the interior of $\mathfrak{A}$ in $\mathbb{T}_f$. Thus, $\mathcal{A}$ is the union of at most a countable collection of classes from $\mathcal{U}$. It follows from Billingsley (1968) and (4.8) that $\mathbf{P}_n$ converges weakly to $\mathbf{P}$. □

Let

$$\mathcal{C}_\mathfrak{t} = \{F \in \mathcal{C}_0\colon \mathfrak{t}(F) = \mathfrak{t}(\mathrm{Int}F)\}.$$

It is easy to show that for each $K \in \mathcal{C}_0$ and $\varepsilon > 0$ there exists a compact $K' \in \mathcal{C}_\mathfrak{t}$ such that $K \subset K' \subset K^\varepsilon$. Thus, the class $\mathcal{C}_\mathfrak{t}$ can be used as a test class in Theorem 4.1.

Proposition 4.3 *The convex compact random set Z_n converges weakly to the RACS $\tilde{Z}$ if*

$$\mathfrak{t}_n(K) \to \tilde{\mathfrak{t}}(K) \quad \text{as } n \to \infty \tag{4.9}$$

for any K from $\mathcal{C}_\mathfrak{t}$.

Similar to Theorem 4.1 the following proposition can be proven. Recall that $\mathcal{H}$ is defined to be the class of all finite intersections of half-spaces.

Proposition 4.4 *The convex compact RACS Z_n converges weakly to the RACS $\tilde{Z}$ if, for any $F \in \mathcal{H}$,*

$$\mathfrak{t}_n(F) \to \tilde{\mathfrak{t}}(F) \quad \text{as } n \to \infty,$$

where $\mathfrak{t}_n, \tilde{\mathfrak{t}}$ are the corresponding inclusion functionals of random sets.

Corollary 4.5 *If the finite dimensional distributions of the support function s_{Z_n} of random convex compact set Z_n, $n \geq 1$, converge to the finite-dimensional distributions of $s_{\tilde{Z}}$, then Z converges weakly to $\tilde{Z}$.*

PROOF. The statement is evident, since the finite-dimensional distributions of s_{Z_n} are determined by the inclusion functional on $\mathcal{H}$. □

Consider now the convergence of expectations of random convex sets. It will be recalled that the (Aumann) *expectation* $\mathbf{E}Z$ of a convex RACS Z is defined as the convex set with the support function $s_{\mathbf{E}Z}(u) = \mathbf{E}s_Z(u)$, $u \in \mathbb{S}^{d-1}$, see Vitale (1988), Stoyan (1989) and Section 2.1.

Theorem 4.6 *Let Z_n, $n \geq 1$, Z be convex random subsets of a certain compact K, and let Z_n converge weakly to Z. Then $\mathbf{E}Z$ converges to $\mathbf{E}Z$ in the Hausdorff metric as $n \to \infty$ and $\mu(\mathbf{E}Z_n) \to \mu(\mathbf{E}Z)$, where μ is the Lebesgue measure in $\mathbb{R}^d$.*

PROOF. It is obvious that $\mathbf{E}s_{Z_n}(u) \to \mathbf{E}s_Z(u)$ as $n \to \infty$ for each u from $\mathbb{S}^{d-1}$. Suppose that

$$\{u_n, n \geq 1\} \subset \mathbb{S}^{d-1}, \quad u_n \to u_0 \text{ as } n \to \infty.$$

By subadditivity of support functions,

$$\begin{aligned} \mathbf{E}s_{Z_n}(u_0) - \mathbf{E}s_{Z_n}(u_0 - u_n) - \mathbf{E}s_Z(u_o) &\leq \mathbf{E}s_{Z_n}(u_n) - \mathbf{E}s_Z(u_0) \\ &\leq \mathbf{E}s_{Z_n}(u_0) + \mathbf{E}s_{Z_n}(u_n - u_0) - \mathbf{E}s_Z(u_0). \end{aligned}$$

Clearly,

$$|\mathbf{E}s_Z(u_0 - u_n)| \leq |s_K(u_0 - u_n)| \to 0 \text{ as } n \to \infty.$$

Hence

$$\sup_{u \in \mathbb{S}^{d-1}} |\mathbf{E}s_{Z_n}(u) - \mathbf{E}s_Z(u)| \to 0 \text{ as } n \to \infty.$$

Thus, $\mathbf{E}Z_n$ converges to $\mathbf{E}Z$ in the Hausdorff metric. The convergence of measures immediately follows from convexity. □

1.5 Set-Theoretic Operations and Measurability.

It should be pointed out that many set-theoretic operations such as union, intersection, Minkowski addition, convex hull etc. preserve measurability of $\mathcal{F}$-valued random elements, provided the result is a closed set. In other words, the result of such operations (if closed) is a random closed set.

First, consider the usual *union* of sets. For any random closed sets X and Y the set $X \cup Y$ is a random closed set too. Its capacity functional is evaluated as

$$T_{X \cup Y}(K) = T_X(K) + T_Y(K) - T_X(K)T_Y(K), \tag{5.1}$$

provided X and Y are independent.

The *Minkowski sum* of X and Y is defined as

$$X \oplus Y = \{x + y \colon x \in X, y \in Y\}. \tag{5.2}$$

If X and Y are independent then the capacity functional of $X \oplus Y$ is evaluated by means of conditional expectations

$$T_{X \oplus Y}(K) = \mathbf{E}\left[\mathbf{E}[T_X(K \oplus \check{Y}) \mid Y]\right],$$

where

$$\check{Y} = \{-y \colon y \in Y\}. \tag{5.3}$$

For single-point random sets the Minkowski addition coincides with the addition of vectors in $\mathbb{R}^d$. Thus, the Minkowski addition for random sets generalizes the usual addition in $\mathbb{R}^d$. On the other hand, the union operation generalizes the maximum scheme for random variables or coordinate-wise maxima for vectors in $\mathbb{R}^d$.

Thus, the Minkowski addition and the union are dual operations in a certain sense, as the addition and the maximum are dual for real numbers. In contrast to its real numbers prototype, the Minkowski addition does not allow to define inverse ones. In other words, $(\mathcal{F}, \oplus)$ is a semi-group only, and $(\mathcal{F}, \cup)$ is a lattice.

The *convex hull* of random sets X and Y is a convex random closed set again. Its inclusion functional is equal to

$$\begin{aligned} \mathfrak{t}(F) &= \mathbf{P}\{\mathrm{conv}(X \cup Y) \subseteq F\} \\ &= \mathbf{P}\{X \subseteq F\}\mathbf{P}\{Y \subseteq F\} = \mathfrak{t}_X(F)\mathfrak{t}_Y(F). \end{aligned} \tag{5.4}$$

Similarly, the *intersection* of X and Y is a random closed set. The *Minkowski subtraction* defined as

$$X \ominus Y = \{x \colon x + Y \subseteq X\} \tag{5.5}$$

is a random closed set too. However, the evaluation of the capacity functionals of $X \cap Y$ and $X \ominus Y$ for general random sets is a very difficult problem.

For convex arguments the Minkowski addition and the convex hull allow a simple translations in terms of support functions (3.3). Namely,

$$s_{X \oplus Y}(u) = s_X(u) + s_Y(u), \tag{5.6}$$

$$s_{\mathrm{conv}(X \cup Y)}(u) = \max(s_X(u), s_Y(u)), u \in \mathbb{S}^{d-1}. \tag{5.7}$$

Thus, the Minkowski addition turns into addition of support functions, and the convex hull turns into the pointwise maximum of support functions of arguments.

For detailed discussions of the above mentioned operations and their properties see Matheron (1975), Serra (1982) and Stoyan et al. (1987).

Note that if X as a random closed set then the *Lebesgue measure* $\mu(X)$, the *norm*

$$\|X\| = \sup\{\|x\| \colon x \in X\},$$

and the extent in a given direction $s_X(u)$ are usual real-valued random variables.

1.6 Regularly Varying Functions.

Similarly to the study of the max-scheme for random variables the analysis of unions of random closed sets involves the technique of regularly varying functions, cf. de Haan (1970), Galambos (1978).

A measurable function $f: \mathbb{R}_+ = [0, \infty) \longrightarrow \mathbb{R}_+$ is *regularly varying* with the *exponent* (or *index*) α if, for each $x > 0$,

$$\lim_{t\to\infty} \frac{f(tx)}{f(t)} = x^\alpha. \tag{6.1}$$

We then write $\mathrm{ind} f = \alpha$. A function L is said to be *slowly varying* if L satisfies (6.1) with $\alpha = 0$, i.e.

$$\lim_{t\to\infty} \frac{L(tx)}{L(t)} = 1, \quad x > 0. \tag{6.2}$$

For backgrounds of the theory of regular varying function see Seneta (1976), de Haan (1970) and Feller (1971).

It is rather easy to see that

$$f(x) = x^\alpha L(x) \tag{6.3}$$

for a certain slowly varying function L. It was proven that (6.2) is valid uniformly for $x \in [a, b]$, $0 < a < b < \infty$. Moreover, a slowly varying function L admits the representation

$$L(x) = \exp\left\{\eta(x) + \int_b^x \frac{\varepsilon(t)}{t} dt\right\}, x \geq b, \tag{6.4}$$

for a certain $b \geq 0$. Here $\eta(x)$ is bounded on $[b, \infty)$ and admits a finite limit as $x \to \infty$, and $\varepsilon(t)$ is a continuous function such that $\varepsilon(t) \to 0$ as $t \to \infty$.

Mention a few properties of regularly varying functions, which will be of use later on. Proofs can be found in Seneta (1976). The letters f and L stand for arbitrary regular varying and slowly varying functions.

1. For each $\gamma > 0$

$$x^\gamma L(x) \to \infty \text{ and } x^{-\gamma} L(x) \to 0 \text{ as } x \to \infty.$$

2. Let $f(x)$ be regularly varying, $\mathrm{ind} f = \gamma$. Then there exists a regularly varying function $\bar{f}(x)$ with exponent $1/\gamma$ such that

$$f(\bar{f}(x)) \sim x \quad \text{as} \quad x \to \infty,$$
$$\bar{f}(f(x)) \sim x \quad \text{as} \quad x \to \infty.$$

Then $\bar{f}$ is said to be the asymptotic inverse function for f.

3. Let L be slowly varying on $[a, \infty)$, $a > 0$, and let $f(x) = x^\gamma L(x)$ be non-decreasing on $[a, \infty)$ for a certain positive γ. Denote

$$\bar{f}(x) = \inf\{y\colon f(y) \geq x, y \in [a, \infty)\}.$$

Then $\bar{f}(x)$ is the asymptotic inverse function for f. This fact is called the *inverse theorem* for univariate regularly varying functions.

For the generalizations to $\mathbb{R}^d$ we follow Yakimiv (1981), see also de Haan and Resnick (1979), de Haan and Omey (1983), Resnick (1986) and references therein.

Let $\mathbb{C}$ be a cone in $\mathbb{R}^d$, and let $S = \mathbb{C} \setminus \{0\}$. A measurable *multivariate* function $f: \mathbb{C} \longrightarrow \mathbb{R}_+$ is said to be *regularly varying* with exponent (or index) α if, for a certain $e \in S$ there exists the finite limit

$$\lim_{t\to\infty} \frac{f(tx)}{f(te)} = \phi(x), \tag{6.5}$$

whatever x belonging to S may be.

The function ϕ is *homogeneous* with exponent α, i.e.

$$\phi(tx) = t^\alpha \phi(x), t > 0, x \in S. \tag{6.6}$$

Following Yakimiv (1981), we then write ind$f = \alpha$, $f \in \Pi_1$ and $\phi \in \mathcal{U}_1$. Clearly, the function ϕ depends on e. If $\mathbb{C} = \mathbb{R}^d$, then e may be safely thought to be can consider e to be equal to the d-tuple $(1, ..., 1)$.

Sometimes (6.5) is too weak to ensure desirable properties of regular varying multivariate functions. The function f is said to belong to Π_2 if (6.5) is valid uniformly on $\mathbb{S}^{d-1} \cap \mathbb{C}$. Namely,

$$\lim_{t\to\infty} \sup_{x \in \mathbb{S}^{d-1}\cap\mathbb{C}} \left| \frac{f(tx)}{f(te)} - \phi(x) \right| = 0. \tag{6.7}$$

It is easy to show that $\Pi_1 = \Pi_2$ if and only if $d = 1$.

Let $\mathfrak{W}_2$ be the class of slowly varying functions on S such that (6.7) is valid with $\phi = 1$. Furthermore, let $\mathcal{U}_2$ be the class of all continuous functions which satisfy (6.6). It was proven in Yakimiv (1981) that $f \in \Pi_2$ if and only if

$$f = L\phi \tag{6.8}$$

for some $\phi \in \mathcal{U}_2$ and $L \in \mathfrak{W}_2$. In fact, this is a multivariate generalization of (6.3).

It was proven in de Haan and Resnick (1987) that for any $L \in \mathfrak{W}_2$ and $c > 0$, $\varepsilon > 0$ there exists t_0 such that for $t \geq t_0$ and $\|x\| \geq c$ it is

$$(1-\varepsilon)\|x\|^{-\varepsilon} \leq \frac{L(tx)}{L(te)} \leq (1+\varepsilon)\|x\|^{\varepsilon}. \tag{6.9}$$

Chapter 2

Survey on Stability of Random Sets and Limit Theorems for Minkowski Addition

2.1 Expectation of a Random Closed Set.

The family $\mathcal{F}$ admits two basic operations. The first one is the union of sets and the second one is the Minkowski addition. In this notes we consider mainly the limit theorems for unions of random sets. Nevertheless, in this chapter a law of large numbers and a limit theorem for Minkowski sums of random sets are shortly discussed. This is explained by the duality between unions and Minkowski sums, which resembles the duality between the addition-scheme and the maximum-scheme for random variables. We discuss only results for random closed subsets of $\mathbb{R}^d$, although very general results for random sets in Banach space are now available, see Gine et al. (1983), Puri and Ralescu (1985), Hiai (1984) and Gine and Hahn (1985a).

First, define the *Aumann expectation* of a random compact set. This notion appeared in Aumann (1965) and later on was extensively used in the theory of multivalued functions and related optimization problems. In context of random sets theory it has been first explored in Artstein and Vitale (1975). In the present section we follow the survey given in Vitale (1988).

Let

$$\|A\| = \sup\{\|x\|\colon\ x \in A\}$$

be the *norm* of a certain random closed set A. It is evident that $\|A\| < \infty$ almost surely if and only if A is compact.

A random vector ξ in $\mathbb{R}^d$ is said to be a *selector* of A if $\xi \in A$ with probability one. The expectation of A is defined to be the set

$$\mathbf{E}A = \{\mathbf{E}\xi\colon\ \xi \text{ is a selector of } A, \mathbf{E}\xi \text{ exists}\}.$$

The condition $\mathbf{E}\|A\| < \infty$ is enough to determine that $\mathbf{E}A$ is non-void and compact. It follows from Aumann (1965) that $\mathbf{E}A$ is convex even for non-convex A. Moreover, $\mathbf{E}A = \mathbf{E}\mathrm{conv}(A)$.

The expectation $\mathbf{E}A$ can be also defined as the convex set having the support function (see Section 1.3)

$$s_{\mathbf{E}A}(u) = \mathbf{E}s_A(u),\ u \in \mathbb{S}^{d-1}.$$

In this way also the expectation of an unbounded random set can be defined.

The following theorem and its various applications are due to Vitale (1988).

Theorem 1.1 (Vitale) *If* $\mathbf{E}\|A\| < \infty$, *then*

$$\mu^{1/d}(\mathbf{E}A) \geq \mathbf{E}\mu^{1/d}(A), \tag{1.1}$$

where μ *is the d-dimensional Lebesgue measure (i.e. volume).*

It was proven in Vitale (1987) that the sequence of convex sets $\mathbf{E}\mathrm{conv}\{\xi_1, \dots, \xi_n\}$, $n \geq 1$, for iid random vectors $\xi_1, \dots, \xi_n$ determines the distribution of ξ_1, provided $\mathbf{E}\|\xi_1\| < \infty$. In turn, the distribution of a random closed set A is determined uniquely by the sequence

$$\mathbf{E}\mathrm{conv}\left\{s_{A_1}(.), \dots, s_{A_n}(.)\right\}, \; n \geq 1,$$

of convex subsets of $C(\mathbb{S}^{d-1})$, where $A_1, A_2, \dots$ are independent copies of A, see Molchanov (1993b).

The definition of the expectation is well-adjusted for the study of Minkowski sums of random sets. The main inconvenience is the necessary convexity of expectations. For instance, the expectation of a stationary random closed set is trivial and equal to $\mathbb{R}^d$.

The variance of A can be also defined to be the set of all variances of its appropriate selectors, see Kruse (1987). However this definition is rather peculiar, since the variance of a non-random set is allowed to be non-trivial. Besides, the variance is very difficult to evaluate.

Further references and discussion on the above mentioned and other notions of expectations, medians and variances of random sets can be found in Stoyan (1989), Stoyan and Stoyan (1992).

2.2 A Strong Law of Large Numbers for Minkowski Sums.

The expectation of a random closed set appears in a *strong law of large numbers* for normalized sums of random closed sets. Its first variant was proven by Artstein and Vitale (1975). Their approach involved two main steps, which also are basic to proving various central limit theorems and laws of the iterated logarithm.

Step 1 Reduce to consideration of random compact convex sets;

Step 2 Prove the result for random compact convex sets by invoking an appropriate result in the space $C(\mathbb{S}^{d-1})$ of continuous functions on $\mathbb{S}^{d-1}$. The appropriate result in $C(\mathbb{S}^{d-1})$ is then applied to the support functions of random sets.

Theorem 2.1 (Artstein and Vitale) *Let* $A, A_1, A_2, \dots$ *be a sequence of independent, identically distributed (iid) random sets with* $\mathbf{E}\|A\| < \infty$. *Then*

$$(A_1 \oplus \cdots \oplus A_n)/n \to \mathbf{E}A \;\; a.s. \; as \; n \to \infty$$

with respect to the Hausdorff metric (or in $\mathbb{T}_k$*).*

The first step of the proof is provided by the following result of Shapley and Folkmann (see Arrow and Hahn, 1971).

Lemma 2.2 (Shapley and Folkmann) *For each $n \geq 1$ and $A_i \in \mathcal{C}_0$, $1 \leq i \leq n$, we have*

$$\rho_{\mathrm{H}}(A_1 \oplus \cdots \oplus A_n, \mathrm{conv}(A_1 \oplus \cdots \oplus A_n)) \leq d^{1/2} \max_{1 \leq i \leq n} \|A_i\|.$$

Then Theorem 2.1 simply follows from the strong law of large numbers in $C(\mathbb{S}^{d-1})$, since the Hausdorff distance between sets is equal to the uniform distance between their support functions. Besides, the norm $\|A_i\|$ coincides with the norm in $C(\mathbb{S}^{d-1})$ of the corresponding support function s_{A_i}.

2.3 A Central Limit Theorem for Random Sets.

The formulation of a *central limit theorem* for random closed sets is more complicated than the corresponding result for random vectors, since Minkowski sums of random sets cannot be centered. It is caused by the lack of an inverse operation to the Minkowski addition. Moreover, $\mathbf{E}A = \{0\}$ yields $A = \{0\}$ with probability one, so that a random set with zero mean is trivial.

Consider a limit theorem for random sets proved by Weil (1982). Let A be a random convex set. Define the *covariance function* of the corresponding support function s_A as

$$\Gamma_A(u,v) = \mathbf{E}\left([s_A(u) - \mathbf{E}s_A(u)][s_A(v) - Es_A(v)]\right),\ u, v \in \mathbb{S}^{d-1}.$$

Let $\xrightarrow{d}$ designates the convergence in distribution of random variables.

Theorem 3.1 (Weil) *Let $A, A_1, A_2, \ldots$ be iid random close sets with $\mathbf{E}\|A\| < \infty$. Then*

$$n^{1/2}\rho_{\mathrm{H}}((A_1 \oplus \cdots \oplus A_n)/n, \mathbf{E}A) \xrightarrow{d} \sup_{u \in \mathbb{S}^{d-1}} \zeta(u) \quad as\ n \to \infty,$$

where $\zeta(u)$ is the Gaussian centered process on $\mathbb{S}^{d-1}$ with the covariance function $\mathbf{E}\zeta(u)\zeta(v) = \Gamma_A(u,v)$. for all u, v from $\mathbb{S}^{d-1}$.

PROOF is also based on the Shapley-Folkman lemma and the central limit theorem in $C(\mathbb{S}^{d-1})$, see Araujo and Gine (1980). □

The particular case of random sets with finite numbers of values was considered in Cressie (1978). It was actually the first limit theorem for random sets.

If the random function $\zeta(u)$ is the support function of a certain random closed set Z, then this set Z is said to be *Gaussian*. It was proven in Lyashenko (1983) and Vitale (1984) that $Z = \xi + M$ for a certain Gaussian vector ξ and a non-random convex compact M. Thus, each Gaussian random set is represented as a Gaussian shift of a non-random compact.

A more intrinsic definition of Gaussian as well as p-stable compact convex sets is due to Gine and Hahn (1985b). A random compact convex set A is called p-stable,

$0 < p \leq 2$, if, for any A_1, A_2 iid with the same law as A and for all $\alpha, \beta \geq 0$, there exist convex closed sets C, D such that

$$\alpha A_1 \oplus \beta A_2 \oplus C \overset{d}{\sim} (\alpha^p + \beta^p)^{1/p} A \oplus D,$$

where $\overset{d}{\sim}$ designates coincidence of probability laws. For $p = 2$ we obtain Gaussian random sets.

If $1 \leq p \leq 2$, then $A = \xi + M$ for a certain p-stable random vector ξ and a non-random convex compact M. If $0 < p < 1$, then A is constructed by means of an appropriate variant of the stochastic integral, see Gine and Hahn (1985b).

2.4 Generalized Expectations of Random Sets

The Aumann expectation of a random set was defined in Section 2.1. It originates deeply in ideas of convex analysis and, for sure, leads to convex-valued results.

Certainly, a relevant expectation of a random set should be set-valued. The Aumann expectation is, moreover, convex-valued. It means that this expectation is always convex, even in the case when the random set in question is of a general nature. This property makes it not very useful for the analysis of non-convex, especially, stationary random closed sets. In the latter case the expectation coincides with the whole space, i.e. it is not informative. A variant of this expectation, proposed by Vitale (1990), also cannot overwhelm its major shortcomings.

On the other hand, a relevant expectation should be closely related with the corresponding law of large numbers, as the Aumann expectation and the strong law of large numbers for the Minkowski addition, see Section 2.2.

Let us consider a more general approach. In special cases it leads to the Aumann expectation or some other expectations discussed in Stoyan (1989), Stoyan and Stoyan (1992). The notion of the expectation depends on a certain family of functions. If this family consists of linear functions only, then we obtain the Aumann expectation. Choosing the class of indicator functions leads to the set of fixed points of a random closed set, see Matheron (1975).

The important tool in the definition of the Aumann expectation is the support function. Clearly, the distribution of a general random closed set is not determined by means of the corresponding support function. For general random sets the definition of the expectation may be based on the corresponding capacity functional $T(K) = \mathbf{P}\{A \cap K \neq \emptyset\}$.

Let $\mathcal{U}$ ($\mathcal{U}_+$) be the family of upper semi-continuous real-valued (positive) functions on E. For any function f from $\mathcal{U}_+$ put

$$\mathbf{E}_A f = \int_0^\infty T(\{u \in E: f(u) \geq t\})dt. \tag{4.1}$$

This capacity integral can be as well defined for a general bounded capacity which is not related to a certain random closed set. Note that if f is upper semicontinuous, then the set $\{u: f(u) \geq t\}$ is closed.

The capacity integral has been applied in Norberg (1985) to the study of the week convergence of random closed sets. It was used also in robust statistics (see Huber, 1981) and the theory of large deviations (see Gerritse, 1983).

The simplest case is when T coincides with a certain probability measure on $\mathbb{R}_+$ (in this case A is a certain singleton $\{\xi\}$ for a positive random variable ξ). Then $\mathbf{E}_A f = \mathbf{E}f(\xi)$, i.e. the integral (4.1) coincides with the expectation of $f(\xi)$. In general it is easy to show that the integral (4.1) is equal to the expectation of the maximum of f over A, i.e.

$$\mathbf{E}_A f = \mathbf{E} \sup_{a \in A} f(a). \tag{4.2}$$

This functional can be defined also for any function f from $\mathcal{U}$, provided the expectation exists. The functional $\mathbf{E}_A f : \mathcal{U} \to \mathbb{R}$ is said to be the *capacity integral.* Sometimes we shall write $f(A)$ instead of $\sup_{a \in A} f(a)$.

If $f(x) = \|x\|$ for each $x \in E$, then $\mathbf{E}_A f = \mathbf{E}\|A\|$. Put $f(x) = (u \cdot x)$ for a certain unit vector u. Then

$$\mathbf{E}_A f = \mathbf{E} \sup_{x \in A} (u \cdot x) = \mathbf{E} s_A(u) = s_{\mathbf{E}A}(u).$$

Furthermore, for the indicator function $f(x) = 1_{x \in K}$ we get $\mathbf{E}_A f = T(K)$.

Consider now some properties of the capacity integral $\mathbf{E}_A f$. First, it is homogeneous, i.e. $\mathbf{E}_A(cf) = c\mathbf{E}_A f$ for each real c. It is easy to show that

$$\mathbf{E}_A \max(f_1, f_2) = \mathbf{E}_A f_1 + \mathbf{E}_A f_2 - \mathbf{E}_A \min(f_1, f_2)$$

if T is a C-additive capacity (see Matheron, 1975) and the functions f_1, f_2 are convex.

Furthermore, if $f(x_1, \ldots, x_n) = x_1 1_{K_1} + x_2 1_{K_2} + \cdots + x_n 1_{K_n}$ for disjoint $K_1, \ldots, K_n$ and $x_1 < x_2 < \ldots < x_n$, then

$$\begin{aligned} \mathbf{E}_A f &= x_1 T(K_1 \cup \ldots \cup K_n) \\ &\quad + (x_2 - x_1) T(K_2 \cup \ldots \cup K_n) + \cdots + (x_n - x_{n-1}) T(K_n). \end{aligned}$$

The capacity integral may be used to define the corresponding set-valued expectation of a random closed set, depending on a certain family $\mathbb{F}$ of numerical functions. Namely, the set

$$\mathbf{E}_{\mathbb{F}} A = \{x : f(x) \le \mathbf{E}_A f, \text{ for all } f \in \mathbb{F}\} \tag{4.3}$$

is said to be $\mathbb{F}$-*expectation* of A.

If $\mathbb{F} \subset \mathcal{U}$, then the set $\mathbf{E}_{\mathbb{F}} A$ is closed. It is evident that $\mathbf{E}_{\mathbb{F}_1} A \subset \mathbf{E}_{\mathbb{F}_2} A$ in the case $\mathbb{F}_2 \subset \mathbb{F}_1$. Similarly, $\mathbf{E}_{\mathbb{F}} A \subset \mathbf{E}_{\mathbb{F}} B$ if $A \subset B$ almost surely.

If the family $\mathbb{F}$ consists of only one function, i.e. $\mathbb{F} = \{f\}$, then $\mathbf{E}_{\mathbb{F}} A$ is equal to $f^{-1}((-\infty, \mathbf{E}_A f])$. Furthermore,

$$\mathbf{E}_{\mathbb{F}_1 \cup \mathbb{F}_2} A = \mathbf{E}_{\mathbb{F}_1} A \cap \mathbf{E}_{\mathbb{F}_2} A,$$

$$\mathbf{E}_{c\mathbb{F}} A = \mathbf{E}_{\mathbb{F}} A, \; c > 0,$$

$$\mathbf{E}_{\mathbb{F}+x} A = \mathbf{E}_{\mathbb{F}} A, \; x \in E,$$

where $c\mathbb{F} = \{cf : f \in \mathbb{F}\}$, $\mathbb{F} + x = \{f + x : f \in \mathbb{F}\}$. Thus, $\mathbf{E}_{\mathbb{F}} A$ is invariant under linear transformations of the family $\mathbb{F}$.

EXAMPLE 4.1 Let $\mathbb{F}$ be the family of indicator functions $1_{x \in K}$ for all compact K belonging to a certain class $\mathfrak{M}$. Then

$$\mathbf{E}_{\mathbb{F}} A = (\cup\{K \in \mathfrak{M} : T(K) < 1\})^c,$$

where M^c denotes the complement to M. If the class $\mathfrak{M}$ is sufficiently rich (c.g., consists of all singletons or all balls) then $\mathbf{E}_{\mathbb{F}}A$ is the set of all fixed points of A, see Matheron (1975) and Section 3.1.

EXAMPLE **4.2** If $\mathbb{F}$ coincides with the class $\mathbb{L}$ of all linear functions, then $\mathbf{E}_{\mathbb{F}}A$ coincides with the Aumann expectation of A.

EXAMPLE **4.3** Let $\mathbb{F}$ be the family of functions

$$f_v(x) = \begin{cases} \|x\| & \text{if } x/\|x\| = v \\ 0 & \text{otherwise} \end{cases}$$

for all $v \in \mathbb{S}^{d-1}$. Then

$$\mathbf{E} \sup_{x \in A} f_v(x) = \mathbf{E} \sup\{r: rv \in A\}.$$

Hence

$$\mathbf{E}_{\mathbb{F}}A = \{x: f_v(x) \leq \mathbf{E} \sup\{r: rv \in A\}\}$$

is the so-called star-shaped expectation of A, see Stoyan (1989).

EXAMPLE **4.4** Let $\mathbb{F}$ be the family of all functions $f_c(x) = c_1 x_1^2 + \cdots + c_d x_d^2$ for $c = (c_1, \ldots, c_d) \in \mathbb{R}^d$, $x = (x_1, \ldots, x_d)$. Then

$$\mathbf{E} \sup_{x \in A} f_c(x) = \mathbf{E} \sup_{x \in A} (c_1 x_1^2 + \cdots + c_d x_d^2) = \mathbf{E} \sup_{x \in A^2} (c_1 x_1 + \cdots + c_d x_d),$$

i.e. $\mathbf{E}_{\mathbb{F}}A$ coincides with the Aumann expectation of the set

$$A^2 = \{(x_1^2, \ldots, x_d^2): (x_1, \ldots, x_d) \in A\}.$$

For example, if $A = \{\xi\} \in \mathbb{R}^d$, then $\mathbf{E}_{\mathbb{F}}A = \{(\mathbf{E}\xi_1^2, \ldots, \mathbf{E}\xi_d^2)\}$.

For any set A put

$$[A]_{\mathbb{F}} = \{x: f(x) \leq \sup\nolimits_{a \in A} f(a), \text{ for all } f \in \mathbb{F}\}.$$

Evidently, $\mathbf{E}_{\mathbb{F}}A = [A]_{\mathbb{F}}$ for any non-random A. A set A is said to be $\mathbb{F}$-*closed* if $[A]_{\mathbb{F}} = A$. It can be easily shown that $\mathbb{F}$-closure is an idempotent operator, i.e. $[[A]_{\mathbb{F}}]_{\mathbb{F}} = [A]_{\mathbb{F}}$.

EXAMPLE **4.5** Let $\mathbb{F} = \mathbb{L}$. Then $\mathbb{F}$-closeness is equivalent to the convexity, and also $[A]_{\mathbb{F}} = \mathrm{conv}(A)$.

Theorem 4.6 *The expectation* $\mathbf{E}_{\mathbb{F}}A$ *is* $\mathbb{F}$-*closed.*

PROOF. Evidently,

$$\begin{aligned} [\mathbf{E}_{\mathbb{F}}A]_{\mathbb{F}} &= \{x: f(x) \leq f(\mathbf{E}_{\mathbb{F}}A), \text{ for all } f \in \mathbb{F}\} \\ &= \{x: f(x) \leq \sup\nolimits_{\{a: f(a) \leq \mathbf{E}_{\mathbb{F}}A, \text{ for all } f \in \mathbb{F}\}} f(a), \text{ for all } f \in \mathbb{F}\} \\ &= \{x: f(x) \leq \mathbf{E}_{\mathbb{F}}A, \text{ for all } f \in \mathbb{F}\} = \mathbf{E}_{\mathbb{F}}A \quad \square \end{aligned}$$

EXAMPLE **4.7** Let $\mathbb{F}$ be the set of all indicators $1_{x\in K}, K \in \mathfrak{M}$. If $\mathfrak{M}$is sufficiently rich (see Example 4.1), then

$$[A]_{\mathbb{F}} = \{x: 1_{x\in K} \leq 1_{A\cap K\neq\emptyset}, K \in \mathfrak{M}\} = \bigcap_{K\cap A, K\in\mathfrak{M}} K^c = \bar{A}$$

EXAMPLE **4.8** Let $A \subset \mathbb{R}$, and let $\mathbb{F}$ be the family of all monotone (increasing) functions. Then

$$[A]_{\mathbb{F}} = (-\infty, \sup A],$$

and

$$\begin{aligned} \mathbf{E}_{\mathbb{F}}A &= \{x: f(x) \leq \mathbf{E}f(\sup A), \text{ for all } f \in \mathbb{F}\} \\ &= (-\infty, \inf_{f\in\mathbb{F}} f^{-1}(\mathbf{E}f(\sup A))), \end{aligned}$$

where f^{-1} is the inverse function to f.

Now consider a *stationary* random closed set A.

Theorem 4.9 *Let A be a stationary random closed set. Suppose that the family $\mathbb{F}$ is translation-invariant, i.e. $\mathbb{F} = \{f(x+v): f \in \mathbb{F}\} = \mathbb{F} \oplus v, v \in \mathbb{R}^d$. Then $\mathbf{E}_{\mathbb{F}}A = \mathbb{R}^d$ or $\mathbf{E}_{\mathbb{F}}A$ is empty.*

PROOF. Since $\mathbb{F}$ is translation-invariant, the stationarity assumption yields

$$\begin{aligned} &\{x: f(x) \leq \mathbf{E}\sup\nolimits_{a\in A} f(a), \text{ for all } f \in \mathbb{F}\} = \\ &\quad = \{x: f(x) \leq \mathbf{E}\sup\nolimits_{a\in A+v} f(a), \text{ for all } f \in \mathbb{F}\} \\ &\quad = \{x: f(x) \leq \mathbf{E}\sup\nolimits_{a\in A} f(a+v), \text{ for all } f \in \mathbb{F}\} \\ &\quad = \{y+v: f(y+v) \leq \mathbf{E}\sup_{a\in A} f(a+v), \text{ for all } f \in \mathbb{F}\} \\ &\quad = \mathbf{E}_{\mathbb{F}\oplus v}A + v = \mathbf{E}_{\mathbb{F}}A + v \end{aligned}$$

for all $v \in \mathbb{R}^d$. Thus, if $\mathbf{E}_{\mathbb{F}}A$ is nonempty, then it coincides with the whole space. □

Therefore, interesting particular examples of $\mathbb{F}$-expectations of stationary random sets can be obtained only in cases where the family $\mathbb{F}$ is not translation-invariant. The following example proposes such a class $\mathbb{F}$, which may be of use in the stationary case.

EXAMPLE **4.10** Let $\mathbb{F}_0$ be a certain class of decreasing functions $f: \mathbb{R}_+ \to \mathbb{R}_+$. For any real-valued random variable ξ put

$$\mathbf{E}_{\mathbb{F}_0}\xi = \{x \geq 0: f(x) \leq \mathbf{E}f(\xi), \text{ for all } f \in \mathbb{F}_0\}.$$

Let $\mathbb{F}(K)$ be the family of functions $f(\rho(x, K))$ for a certain compact K and f running through the family $\mathbb{F}_0$. Here $\rho(A, B)$ is the minimal distance between the points of A and B. Then the $\mathbb{F}$(K)-expectation of A

$$\begin{aligned} \mathbf{E}_{\mathbb{F}(K)}A &= \{x: \rho(x, K) \in \mathbf{E}_{\mathbb{F}_0}\rho(A, K)\} \\ &= \{x: f(\rho(x, K)) \leq \mathbf{E}f(\rho(A, K)), \text{ for all } f \in \mathbb{F}_0\}. \end{aligned}$$

For example, if $\mathbb{F}_0$ consists of only linear functions, then $\mathbf{E}_{\mathbb{F}_0}\xi = [0, \mathbf{E}\xi]$, and

$$\mathbf{E}_{\mathbb{F}(K)}A = K \oplus b(o, \mathbf{E}\rho(A,K)) = K^{\mathbf{E}\rho(A,K)}.$$

Let A be the stationary Poisson point process with intensity λ. Then

$$\mathbf{P}\{\rho(A,K) \leq x\} = 1 - \exp\{-\lambda\mu_d(K^x)\},$$

where μ_d is the d-dimensional Lebesgue measure. If K is convex, then the expectation $\mathbf{E}\rho(A,K)$ can be expressed in terms of the corresponding Minkowski functionals of K, since $\mu_d(K^x)$ can be evaluated by means of the Steiner formula, see Matheron (1975). Hence in this scheme we obtain a family of expectations, depending on K. E.g., let $d = 1$, and let $K = [0, h]$. Then

$$\mathbf{P}\{\rho(A,K) \leq x\} = 1 - \exp\{-\lambda(h + 2x)\}.$$

Hence $\mathbf{E}\rho(A,K) = e^{-\lambda h}/(2\lambda) = g(h)$. Thus

$$\mathbf{E}_{\mathbb{F}(K)}A = [-g(h), h + g(h)].$$

Other interesting examples may be obtained if the class $\mathbb{F}_0$ consists of the functions $e^{-\alpha x}$ for different $\alpha > 0$.

Now reformulate the definition of the $\mathbb{F}$-expectation by means of selectors. The selector variant of the $\mathbb{F}$-expectation is defined as

$$\mathbf{E}^s_{\mathbb{F}}A = \{\mathbf{E}_{\mathbb{F}}\{\xi\} : \xi \in \mathcal{S}(A)\}$$

where $\mathcal{S}(A)$ is the family of all selectors of A (of course, $\mathbf{E}^s_{\mathbb{F}}A$ is empty if $\mathbf{E}_{\mathbb{F}}\{\xi\}$ is empty for all $\xi \in \mathcal{S}(A)$). In general, $\mathbf{E}^s_{\mathbb{F}}A$ is not $\mathbb{F}$-closed. For example, if the family $\mathbb{F}$ coincides with the family of all linear functions, then the $\mathbb{F}$-closeness is equivalent to convexity, but the convexity property of $\mathbf{E}^s_{\mathbb{F}}A$ depends on the (atomic or non-atomic) structure of the probability space in question, see Vitale (1990).

Unfortunately, the $\mathbb{F}$-expectation does not always coincide with its selector variant.

Theorem 4.11 *Suppose that for any random vector ξ in $\mathbb{R}^d$*

$$\sup\{f(u) : u \in \mathbf{E}_{\mathbb{F}}\xi\} = \mathbf{E}f(\xi), f \in \mathbb{F} \tag{4.4}$$

Then $\mathbf{E}_{\mathbb{F}}A = [\mathbf{E}^s_{\mathbb{F}}A]_{\mathbb{F}}$, i.e. $\mathbf{E}_{\mathbb{F}}A$ is the $\mathbb{F}$-closure of the family of $\mathbb{F}$-expectations of all selectors.

PROOF. Evidently, $\mathbf{E}^s_{\mathbb{F}}A \subset \mathbf{E}_{\mathbb{F}}A$. Moreover, $[\mathbf{E}^s_{\mathbb{F}}A]_{\mathbb{F}} \subset \mathbf{E}_{\mathbb{F}}A$, since $\mathbf{E}_{\mathbb{F}}A$ is $\mathbb{F}$-closed. Furthermore,

$$[\mathbf{E}^s_{\mathbb{F}}A]_{\mathbb{F}} = \bigcap_{f\in\mathbb{F}}\{x : f(x) \leq \sup_{a\in\mathbf{E}^s_{\mathbb{F}}A} f(a)\} = \bigcap_{f\in\mathbb{F}}\{x : f(x) \leq \sup_{\xi\in\mathcal{S}(A)} f(\mathbf{E}_{\mathbb{F}}\xi)\}$$

where $f(\mathbf{E}_{\mathbb{F}}\xi) = \sup\{f(u) : u \in \mathbf{E}_{\mathbb{F}}\xi\} = \mathbf{E}f(\xi)$ by the condition of Theorem. Thus,

$$[\mathbf{E}^s_{\mathbb{F}}A]_{\mathbb{F}} = \bigcap_{f\in\mathbb{F}}\{x : f(x) \leq \sup_{\xi\in\mathcal{S}(A)} \mathbf{E}f(\xi)\} = \bigcap_{f\in\mathbb{F}}\{x : f(x) \leq \mathbf{E}f(A)\} = \mathbf{E}_{\mathbb{F}}A.$$

Suppose that $x \in \mathbf{E}_{\mathbb{F}}A \setminus [\mathbf{E}^s_{\mathbb{F}}A]_{\mathbb{F}}$. Then there exists a function f such that $f(x) > \mathbf{E}f(\xi)$ for all $\xi \in \mathcal{S}(A)$. On the other hand, $f(x) \leq \mathbf{E}f(A)$. Hence $\mathbf{E}f(A) > \mathbf{E}f(\xi)$ for all $\xi \in \mathcal{S}(A)$, contrary to the fact that $\sup_{a\in A} f(x) = f(\eta)$ for a certain selector $\eta \in \mathcal{S}(A)$. $\square$

Corollary 4.12 *If (4.4) is valid, then for each $f \in \mathbb{F}$*

$$f(\mathbf{E}_{\mathbb{F}}A) = \mathbf{E}_A f. \tag{4.5}$$

PROOF. Since $\mathbf{E}_{\mathbb{F}}A = [\mathbf{E}^s_{\mathbb{F}}A]_{\mathbb{F}}$, we get

$$f(\mathbf{E}_{\mathbb{F}}A) = f(\mathbf{E}^s_{\mathbb{F}}A) = \sup_{\xi\in\mathcal{S}(A)} f(\mathbf{E}_{\mathbb{F}}\xi) = \sup_{\xi\in\mathcal{S}(A)} \mathbf{E}_f(\xi) = \mathbf{E}_A f. \quad \square$$

For example, the condition (4.4) is satisfied for the family of linear functions and also for the functions of Example 4.3.

Similarly to the Aumann expectation, which comes over in the law of large numbers for Minkowski addition, $\mathbb{F}$-expectation appears in the strong law of large numbers for a special addition defined by means of the family $\mathbb{F}$.

For each A, B put

$$[A, B]_{\mathbb{F}} = \{x: f(x) \le f(A) + f(B), \text{ for all } f \in \mathbb{F}\}.$$

Recall that $f(A)$ denotes $\sup_{a\in A} f(a)$.

If $\mathbb{F} = \mathbb{L}$, then $[A, B]_{\mathbb{F}}$ is the convex hull of the Minkowski sum $A \oplus B$. For the family $\mathbb{F}$ of all indicator functions of all compacts we get

$$\begin{aligned} [A, B]_{\mathbb{F}} &= \{x: 1_{x\in K} \le 1_{A\cap K\neq\emptyset} + 1_{B\cap K\neq\emptyset}, K \in \mathcal{K}\} \\ &= \bigcap_{K\cap(A\cup B), K\in\mathcal{K}} K^c = \overline{A \cup B}. \end{aligned}$$

If $\mathbb{F}$ is the family of indicators of convex sets, then $[A, B]_{\mathbb{F}}$ is the convex hull of the union of A and B. If $\mathbb{F}$ is the family of functions $\{cx^2: c \ge 0\}$ on the real line then for each point pair x, y their $\mathbb{F}$-sum is $(x^2 + y^2)^{1/2}$.

For all non-random A, B and $f \in \mathbb{F}$ suppose that

$$f([A, B]_{\mathbb{F}}) = f(A) + f(B), \tag{4.6}$$

i.e.

$$\sup\{f(x): g(x) \le g(A) + g(B), \text{ for all } g \in \mathbb{F}\} = f(A) + f(B)$$

(in general, the left-hand side is not greater than the right-hand one). Then $\mathbb{F}$-addition has especially good properties.

Theorem 4.13 *If (4.6) is valid, then the $\mathbb{F}$-addition is associative, i.e.*

$$[[A, B]_{\mathbb{F}}, C]_{\mathbb{F}} = [A, [B, C]_{\mathbb{F}}]_{\mathbb{F}}.$$

PROOF. It follows from (4.6) that

$$\begin{aligned} [[A, B]_{\mathbb{F}}, C]_{\mathbb{F}} &= \{x: f(x) \le f([A, B]_{\mathbb{F}}) + f(C), \text{ for all } f \in \mathbb{F}\} \\ &= \{x: f(x) \le f(A) + f(B) + f(C), \text{ for all } f \in \mathbb{F}\} \\ &= [A, [B, C]_{\mathbb{F}}]_{\mathbb{F}}. \quad \square \end{aligned}$$

Then we can write $[A, B, C]_{\mathbb{F}}$ instead of $[[A, B]_{\mathbb{F}}, C]_{\mathbb{F}}$. The condition (4.4) implies also the $\mathbb{F}$-additivity of the $\mathbb{F}$-expectation.

Theorem 4.14 *It (4.4) and (4.6) are valid and* $\mathbf{E}_{\mathbb{F}}A$ *and* $\mathbf{E}_{\mathbb{F}}B$ *are non-void, then*

$$\mathbf{E}_{\mathbb{F}}[A, B]_{\mathbb{F}} = [\mathbf{E}_{\mathbb{F}}A\,, \mathbf{E}_{\mathbb{F}}B]_{\mathbb{F}}.$$

PROOF. Formula (4.6) implies

$$\begin{aligned} \mathbf{E}_{\mathbb{F}}[A, B]_{\mathbb{F}} &= \{x: f(x) \leq \mathbf{E}f([A, B]_{\mathbb{F}}), \text{ for all } f \in \mathbb{F}\} \\ &= \{x: f(x) \leq \mathbf{E}f(A) + \mathbf{E}f(B), \text{ for all } f \in \mathbb{F}\}\,. \end{aligned}$$

On the other hand, Corollary 4.12 yields

$$\begin{aligned} [\mathbf{E}_{\mathbb{F}}A\,, \mathbf{E}_{\mathbb{F}}B]_{\mathbb{F}} &= \{x: f(x) \leq f(\mathbf{E}_{\mathbb{F}}A\,) + f(\mathbf{E}_{\mathbb{F}}B), \text{ for all } f \in \mathbb{F}\} \\ &= \{x: f(x) \leq \mathbf{E}f(A) + \mathbf{E}f(B), \text{ for all } f \in \mathbb{F}\}\,. \quad \square \end{aligned}$$

For each nonnegative c put

$$c \circ A = \{x: f(x) \leq cf(A), \text{ for all } f \in \mathbb{F}\}\,.$$

Now consider the normed sums and the corresponding law of large numbers. For each n and iid random sets $A_1, \ldots, A_n$ put

$$\frac{1}{n} \circ [A_1, \ldots, A_n]_{\mathbb{F}} = \{x: f(x) \leq \tfrac{1}{n} \textstyle\sum_{i=1}^{n} f(A_i), \text{ for all } f \in \mathbb{F}\}\,. \tag{4.7}$$

If $\mathbb{F} = \mathbb{L}$, then the left-hand side is equal to $\frac{1}{n}(A_1 \oplus \cdots \oplus A_n)$. For the family of functions from Example 4.3 the set (4.7) coincides with star-shaped averages, considered in Stoyan (1989).

The left-hand side can be defined also for a family $\mathbb{F}$, which does not satisfy (4.6). If $\mathbb{F}$ is the family of all indicators of compact sets, then (4.7) is equal to the closure of $(A_1 \cap \cdots \cap A_n)$.

Define iid random functionals on $\mathbb{F}$ as $\xi_i(f) = f(A_i)$. Then

$$\frac{1}{n} \circ [A_1, \ldots, A_n]_{\mathbb{F}} = \{x: f(x) \leq \zeta_n(f), \text{ for all } f \in \mathbb{F}\}\,,$$

where

$$\zeta_n(f) = \frac{1}{n} \sum_{i=1}^{n} \xi_i(f).$$

For each f the strong law of large numbers yields $\zeta_n(f) \to \mathbf{E}f(A) = \mathbf{E}\xi_i(f)$ a.s. as $n \to \infty$.

Let us now introduce also a metric on the class of sets called the $\mathbb{F}$-metric. Note first that each $\mathbb{F}$-closed set A can be obtained as

$$A = \{x: f(x) \leq f(A), \text{ for all } f \in \mathbb{F}\}\,.$$

For $\varepsilon > 0$ put

$$[A]_{\mathbb{F}}^{\varepsilon} = \{x: f(x) \leq f(A) + \varepsilon, \text{ for all } f \in \mathbb{F}\}\,.$$

Certainly,

$$\bigcap_{\varepsilon>0}[A]^{\varepsilon}_{\mathbb{F}} = [A]_{\mathbb{F}}.$$

If $\mathbb{F} = \mathbb{L}$, then $[A]^{\varepsilon}_{\mathbb{F}}$ is equal to the Minkowski sum $\mathrm{conv}(A) \oplus b(o,\varepsilon)$.

Define the distance between $\mathbb{F}$-closed sets A and B as

$$\rho_{\mathbb{F}}(A,B) = \inf\{\varepsilon > 0 \colon A \subset [B]^{\varepsilon}_{\mathbb{F}}, B \subset [A]^{\varepsilon}_{\mathbb{F}}\}.$$

If $\mathbb{F} = \mathbb{L}$, then $\rho_{\mathbb{F}}$ coincides with the Hausdorff distance between the convex hulls of A and B. If $\mathbb{F}$ is the family of indicators of compact sets, then $[A]^{\varepsilon}_{\mathbb{F}}$ is the closure of A in case $\varepsilon < 1$ and $[A]^{\varepsilon}_{\mathbb{F}} = \mathbb{R}^d$ if $\varepsilon \geq 1$. Hence $\rho_{\mathbb{F}}(A,B) = 1$ for $A \neq B$.

It is evident that $A \subset [B]^{\varepsilon}_{\mathbb{F}}$ iff $f(x) \leq f(B) + \varepsilon$ for all $x \in A$ and $f \in \mathbb{F}$. In other words, $f(A) \leq f(B) + \varepsilon$ for all $f \in \mathbb{F}$. Therefore,

$$\begin{aligned}\rho_{\mathbb{F}}(A,B) &= \inf\{\varepsilon > 0 \colon f(A) \leq f(B) + \varepsilon, f(B) \leq f(A) + \varepsilon, \text{ for all } f \in \mathbb{F}\} \\ &= \sup_{f \in \mathbb{F}} |f(A) - f(B)|\end{aligned}$$

In particular, for $\mathbb{F} = \mathbb{L}$ we obtain the fact that the Hausdorff distance between convex sets is equal to the uniform distance between their support functions.

Theorem 4.15 *The set $B = \mathbf{E}_{\mathbb{F}}A$ gives the minimum to the functional $\mathbf{E}(\rho_{\mathbb{F}}(B,A))^2$ for B belonging to the family $\mathcal{F}_{\mathbb{F}}$ of $\mathbb{F}$-closed sets.*

PROOF. Clearly,

$$\mathbf{E}\rho_{\mathbb{F}}(B,A)^2 = \mathbf{E} \sup_{f \in \mathbb{F}} (f(B) - f(A))^2.$$

Then for each $B \in \mathcal{F}_{\mathbb{F}}$ the distance $\rho_{\mathbb{F}}(B,A)$ coincides with the uniform distance between two functionals $\phi_A, \phi_B \colon \mathcal{F}_{\mathbb{F}} \to \mathbb{R}$, where $\phi_A(f) = f(A), \phi_B(f) = f(B)$. Then the value

$$\mathbf{E}\rho_{\mathbb{F}}(B,A)^2 = \mathbf{E} \sup_{f \in \mathbb{F}} (\phi_A(f) - \phi_B(f))^2$$

is minimal if $\phi_B(f) = \mathbf{E}\phi_A(f), f \in \mathbb{F}$. Thus, $f(B) = \mathbf{E}f(A)$ for each $f \in \mathbb{F}$. Since B is $\mathbb{F}$-closed, we get

$$\begin{aligned}B &= \{x \colon f(x) \leq f(B), \text{ for all } f \in \mathbb{F}\} \\ &= \{x \colon f(x) \leq \mathbf{E}f(A), \text{ for all } f \in \mathbb{F}\} = \mathbf{E}_{\mathbb{F}}A. \quad \square\end{aligned}$$

The following theorem is a strong law of large numbers for normalized $\mathbb{F}$-sums of random closed sets.

Theorem 4.16 *If $\zeta_n(f) \to \mathbf{E}f(A) = \mathbf{E}\xi_i(f)$ a.s. as $n \to \infty$ uniformly for all $f \in \mathbb{F}$ (i.e. the functional $\xi_i(f)$ satisfies the Glivenko-Cantelli theorem on the class $\mathbb{F}$), then*

$$\rho_{\mathbb{F}}(\frac{1}{n} \circ [A_1, \ldots, A_n]_{\mathbb{F}}, \mathbf{E}_{\mathbb{F}}A) \to 0 \quad a.s. \text{ as } n \to \infty. \tag{4.8}$$

Similarly a central limit theorem in the $\rho_{\mathbb{F}}$-metric can be derived from the central limit theorem for the functionals on the class $\mathbb{F}$, cf Weil (1982).

Now we shall find some sufficient conditions ensuring the uniform convergence of the functional $\zeta_n(f)$ to $\mathbf{E}f(A)$ for all $f \in \mathbb{F}$. Let $\mathbb{F}$ be equipped with a certain topology $\mathcal{T}_{\mathbb{F}}$.

Theorem 4.17 *Suppose that $(\mathbb{F}, \mathcal{T}_{\mathbb{F}})$ is a certain compact space, and for any $f \in \mathbb{F}$ and a family of its neighborhoods $U(f)$ it is*

$$\mathbf{E}\left[\sup_{g \in U(f)} g(A) - \inf_{g \in U(f)} g(A)\right] \to 0 \quad as \quad U(f) \downarrow \{f\}. \tag{4.9}$$

Then

$$\sup_{f \in \mathbb{F}} |\zeta_n(f) - \mathbf{E}f(A)| \to 0 \quad a.s. \quad as \quad n \to \infty,$$

and (4.8) is valid.

PROOF. The proof follows the standard scheme, given in Bhattacharia and Ranga Rao (1976) for general probability measures. □

Suppose that the convergence in $\mathbb{F}$ is compatible with the $\rho_{\mathbb{F}}$-metric in such a way that for any $f \in \mathbb{F}$, its neighborhood $U(f)$ in $\mathcal{T}_{\mathbb{F}}$, and $A \in \mathcal{F}_{\mathbb{F}}$

$$f([A]_{\mathbb{F}}^{-\varepsilon}) \le \inf_{g \in U(f)} g(A) \le \sup_{g \in U(f)} g(A) \le f([A]_{\mathbb{F}}^{\varepsilon}) \tag{4.10}$$

for a certain $\varepsilon > 0$, where

$$[A]_{\mathbb{F}}^{-\varepsilon} = \{x \colon [\{x\}]_{\mathbb{F}}^{\varepsilon} \subset A\}.$$

The $\mathbb{F}$-interior of A is defined as

$$\mathrm{Int}_{\mathbb{F}} A = \bigcup_{\varepsilon > 0} [A]_{\mathbb{F}}^{-\varepsilon}.$$

The set A is said to be a.s. $\mathbb{F}$-canonically closed if

$$[\mathrm{Int}_{\mathbb{F}} A]_{\mathbb{F}} = A \text{ a.s.},$$

i.e. A coincides almost surely with the $\mathbb{F}$-closure of its $\mathbb{F}$-interior, cf Molchanov (1987). Then Theorem 4.17 can be reformulated as follows.

Theorem 4.18 *Suppose that the space $(\mathbb{F}, \mathcal{T}_{\mathbb{F}})$ is compact and for each $f \in \mathbb{F}$*

$$\mathbf{E}f(A) = \mathbf{E}f(\mathrm{Int}_{\mathbb{F}} A).$$

If A is a.s. $\mathbb{F}$-canonically closed, then $\zeta_n(f)$ converges to $\mathbf{E}f(A)$ uniformly for $f \in \mathbb{F}$, i.e. (4.8) is valid.

PROOF. First combine (4.10) and (4.9). Then use the monotone convergence theorem for expectations. □

The functional ξ satisfies the Glivenko-Cantelli theorem on $\mathbb{F}$ for each random set A if the class $\mathbb{F}$ is a so-called Vapnik-Chervonenkis class (VC-class) of functions, see Dudley (1984). The family $\mathbb{L}$ is a VC-class, so that this theorem implies the strong law of large numbers for the Aumann expectation. If $\mathbb{F}$ is the class of functions of Example 4.3, then we obtain the law of large numbers for the star-shaped expectation.

If $\mathbb{F}$ is the class of indicators of compact sets then $\zeta_n(f)$ is the empirical capacity, defined for iid observations of the random set A, see Molchanov (1987), where necessary and sufficient conditions for the uniform convergence were found.

Chapter 3

Infinite Divisibility and Stability of Random Sets with respect to Unions

3.1 Union-Stable Random Sets.

The first to study the stability of random closed sets with respect to their unions (U-stability) was Matheron (1975). He characterized union-infinitely-divisible random sets and considered the simplest case of union-stable sets (without fixed points). These notions were later discussed from a very general point of view by Trader (1981). However Trader's constructions evaded some difficulties. For instance, the characterization problem was merely reduced to some functional equations.

A random closed set A is said to be *infinitely divisible for unions* if, for any positive integer n,

$$A \overset{d}{\sim} A_{n1} \cup \cdots \cup A_{nn},$$

where A_{ni}, $1 \leq i \leq n$, are iid random closed sets, see Matheron (1975). Hereafter $\overset{d}{\sim}$ designates equivalence in distributions.

To proceed further the notion of a *fixed point* should be introduced. The point x is said to be a fixed point of A if

$$\mathbf{P}\{x \in A\} = T(\{x\}) = 1,$$

where T is the capacity functional of A. In other words, x is a fixed point if and only if A contains x almost surely. The *set of all fixed points* of A is denoted by F_A.

EXAMPLE 1.1 Let $A = (-\infty, \xi]$ be a random subset of $\mathbb{R}^1$. If the random variable ξ is a.s. positive, then $F_A \supseteq (-\infty, 0]$.

The random closed set A is said to be *non-trivial* if $\mathbf{P}\{A = F_A\} < 1$, i.e. A does not coincide almost surely with the set of its fixed points.

Clearly, $T(K) = 1$ as soon as K hits F_A. To exclude such a case introduce the class

$$\mathcal{K}_A = \{K \in \mathcal{K}\colon\ K \cap F_A = \emptyset\}.$$

It is easy to prove that F_A is a closed set. Having replaced $\mathbb{R}^d$ by the space $\mathbb{R}^d \setminus F_A$, we can consider only union-infinitely-divisible random sets without fixed points, as it was shown in Matheron (1975). The following theorem provides a slight modification of his result, see also Norberg (1984). The similar result can be obtained by the instrumentality of the harmonic analysis on semigroups, see Berg, Christensen and Ressel (1984).

Theorem 1.2 *The RACS A is union-infinitely-divisible if and only if its capacity functional is represented as*

$$T(K) = 1 - \exp\{-\Psi(K)\}, \tag{1.1}$$

where $\Psi(K)$ is an alternating Choquet capacity of infinite order such that $\Psi(\emptyset) = 0$ and $\Psi(K)$ is finite for all K belonging to $\mathcal{K}_A$.

Union-stable sets form a sub-class of union-infinitely-divisible random closed sets. A random closed set A is said to be *union-stable* (U-stable) if, for any $n > 0$,

$$a_n A \overset{d}{\sim} A_1 \cup \dots \cup A_n, \tag{1.2}$$

where $a_n > 0$ and $A_1, ..., A_n$ are independent copies of A.

The notion of U-stable random closed set generalizes the famous definition of max-stable random variables, see Galambos (1978) and Leadbetter et al. (1986). First, recall several facts from the theory of extremes.

A random variable ξ is said to be *max-stable* if, for all $n > 1$,

$$a_n \xi + b_n \overset{d}{\sim} \max(\xi_1, \dots, \xi_n),$$

where $a_n > 0$, $b_n \in \mathbb{R}^d$, and $\xi_1, \dots, \xi_n$ are iid copies of ξ. It is well-known (see, e.g., Galambos, 1978) that up to a shift any nondegenerate max-stable distribution function is of type one and only one distribution of the parametric family

$$F_\gamma(x) = \exp\left\{-(1+\gamma x)^{-1/\gamma}\right\}, \gamma x \geq -1, \gamma \in \mathbb{R}. \tag{1.3}$$

Besides, if $\gamma > 0$ (type I), then $F_\gamma(x) = 0$ for $x \leq -1/\gamma$, if $\gamma < 0$ (type II) then $F_\gamma(x) = 1$ for $x \geq -1/\gamma$, and if $\gamma = 0$ (type III), then $(1+\gamma x)^{-1/\gamma}$ is an abuse of language for e^{-x}.

Max-stable vectors in $\mathbb{R}^d$ were studied by Balkema and Resnick (1977), de Haan and Resnick (1977). Max-stable random processes were considered in de Haan (1984), and from the very general point of view in Norberg (1986a, 1987), see also Gine et al. (1990). There are close connections between max-stable random processes and union-stable random sets, since the hypograph of any max-stable process is a union-stable random set (cf. Section 8.3).

Union-stable sets without fixed points were characterized in Matheron (1975). He proved that the capacity Ψ from (1.1) is homogeneous with the positive exponent α if and only if A is U-stable and $F_A = \emptyset$. It should be noted that the general situation cannot be reduced to the case $F_A = \emptyset$ by considering the space $\mathbb{R}^d \setminus F_A$, since $\mathbb{R}^d \setminus F_A$ is not a closed cone any longer. On the other hand, the following example shows that even in $\mathbb{R}^1$ there are simple examples of U-stable sets which do not fall in with Matheron's scheme.

EXAMPLE 1.3 Let $A = (-\infty, \xi]$ be a random subset of $\mathbb{R}^1$. It is evident that A is U-stable iff ξ is a max-stable random variable. Matheron's characterization theorem yields

$$1 - T(K) = \mathbf{P}\{\xi < \inf K\} = \begin{cases} \exp\{-c(-x)^\alpha\} & , \quad x < 0 \\ 1 & , \quad x \geq 0 \end{cases}, x = \inf K,$$

where $\alpha > 0$ and c is a certain positive constant. Thus, only max-stable laws of type II can be characterized. It is evident that max-stable laws of type I cannot be obtained from Matheron's characterization theorem, since ξ is positive almost surely, so that $F_A = (-\infty, 0]$ is non-empty.

Although the proof in Matheron (1975) was relied essentially on the lack of fixed points, a similar characterization theorem is valid for general U-stable random sets, see Molchanov (1992). It should be noted that the characterization of union-stable random sets is more difficult than the characterization of max-stable random variables. The main difficulty is caused by the possible self-similarity of random sets. Namely, if $\xi \overset{d}{\sim} c\xi$ for a random variable ξ and each $c > 0$, then ξ is equal to 1 a.s. On the other hand, the relation $A \overset{d}{\sim} cA$ for a random set A admits a lot of solutions, say the set of zeros of the Wiener process.

Theorem 1.4 *A non-trivial random closed set A is U-stable iff its capacity functional T is of the form (1.1), where $\Psi(K)$ is a Choquet capacity, $\Psi(\emptyset) = 0$ and*

$$\Psi(sK) = s^\alpha \Psi(K), \Psi(K) < \infty, \tag{1.4}$$

$$sF_A = F_A \tag{1.5}$$

for a certain $\alpha \neq 0$, whatever positive s and K from $\mathcal{K}_A$ may be.

PROOF of the *necessity* falls into several stages.

I. Let $T(aK) = T(a_1 K)$ for all K from $\mathcal{K}$ and certain $a, a_1 > 0$. Prove that $a = a_1$. It is sufficient to consider the case $a_1 = 1$, $a < 1$. Then, for any $n \geq 1$,

$$T(K) = T(a^n K), \ K \in \mathcal{K}.$$

Hence $T(K) \leq T(B_\varepsilon(0))$ for each $\varepsilon > 0$. Semi-continuity of T implies $T(K) \leq T(\{0\})$, $K \in \mathcal{K}$. Thus, $T(\{0\}) \geq T(\mathbb{R}^d) > 0$, since A is non-empty with positive probability. It follows from (1.2) that

$$T(\{0\}) = 1 - (1 - T(\{0\}))^n,$$

so that $T(\{0\}) = 1$, whence $0 \in F_A$. If $F_A = \emptyset$, then the first step has been proven.

Let F_A be non-void. The condition (1.2) yields

$$T(K) = 1 - (1 - T(a_n K))^n, n \geq 1, K \in \mathcal{K} \tag{1.6}$$

for some $a_n > 0$. Since $a \neq 1$

$$a_n = a^{m(n)} \Delta_n, n \geq 1,$$

for a certain integer $m(n)$ and Δ_n belonging to $(a,1]$. Then, for each compact K and $n \geq 1$,

$$T(K) = 1 - \left(1 - T(a_n K / a^{m(n)})\right)^n = 1 - (1 - T(\Delta_n K))^n. \tag{1.7}$$

Choose $\varepsilon > 0$ such that $F_A^\varepsilon \neq \mathbb{R}^d$ and denote

$$K_\varepsilon = \overline{\mathbb{R}^d \setminus F_A^\varepsilon} \cap B_R(0).$$

Then for sufficiently small ε and large R

$$0 < T(K_\varepsilon) < 1.$$

It follows from (1.7) that

$$T(K_\varepsilon) = 1 - (1 - T(\Delta_n K_\varepsilon))^n.$$

Then

$$T(\Delta_n K_\varepsilon) \to 0 \ \text{ as } \ n \to \infty.$$

It is obvious that

$$\begin{aligned} T(\Delta_n K_\varepsilon) &= \mathbf{P}\{A \cap \Delta_n K_\varepsilon \neq \emptyset\} \\ &= 1 - \mathbf{P}\{A \subset (\Delta_n K_\varepsilon)^c\} \\ &= 1 - \mathbf{P}\left\{A \subset \Delta_n (F_A^{\varepsilon -} \cup B_R(0))\right\}, \end{aligned}$$

where $F_A^{\varepsilon -} = \operatorname{Int} F_A^\varepsilon$. Without loss of generality suppose $\Delta_n \to \Delta \in [a,1]$ as $n \to \infty$. Hence

$$\mathbf{P}\left\{A \subset (\Delta F_A)^{\varepsilon + \delta_n} \cup B_{R_1}^c(0)\right\} \to 1 \ \text{ as } \ n \to \infty,$$

where $\delta_n \downarrow 0$ as $n \to \infty$, $R_1 = \Delta(R - \delta_0)$, $\delta_0 > 0$. Thus

$$\mathbf{P}\left\{A \subset (\Delta F_A)^\varepsilon \cup B_{R_1}^c(0)\right\} = 1.$$

Letting ε go to zero and R go to infinity yields

$$\mathbf{P}\{A \cap K \subseteq \Delta F_A\} = 1$$

for any K from $\mathcal{K}$. Hence $A \subseteq \Delta F_A$ almost surely. It is easy to derive from (1.7) that $\Delta_n F_A = F_A$. Thus, $\Delta F_A = F_A$ and $A \subseteq F_A$ with probability one, whence A is trivial, contrary to the condition of Theorem. Thus, $a = 1$.

II. Since a U-stable set is union-infinitely-divisible, its capacity functional has the form (1.1). It follows from (1.2) that

$$\begin{aligned} n\Psi(a_n K) &= \Psi(K),\ n \geq 1,\ K \in \mathcal{K}_A, \\ a_n F_A &= F_A. \end{aligned}$$

For any positive rational number $s = m/n \in \mathbb{Q}_+$ put $a(s) = a_m / a_n$. It is easy to show that $a(s)$ does not depend on the representation of s. Then, for any s from $\mathbb{Q}$,

$$\begin{aligned} s\Psi(a(s)K) &= \Psi(K),\ s > 0,\ K \in \mathcal{K}_A, \\ a(s) F_A &= F_A. \end{aligned} \tag{1.8}$$

III. Let s, s_1 belong to $\mathbb{Q}_+$. It follows from (1.8) that

$$\Psi(a(s)a(s_1)K) = \Psi(a(ss_1)K).$$

The first step of the proof and (1.1) yield

$$a(ss_1) = a(s)a(s_1). \tag{1.9}$$

IV. Prove that $a(s_n) \to 1$ as $s_n \to 1$, i.e. the function $a(s)$ is continuous on $\mathbb{Q}_+$ at $s = 1$. It follows from (1.1), (1.8) that

$$T(a(s_n)K) \to T(K) \text{ as } n \to \infty, K \in \mathcal{K}.$$

Without loss of generality suppose that the sequence $a(s_n)$, $n \geq 1$, has the limit (which is allowed to be infinite).

Let this limit be finite and equal to $a > 0$. Then, for any $\varepsilon > 0$ and sufficiently large n,

$$T(a(s_n)K) \leq T(aK^\varepsilon).$$

Hence $T(K) \leq T(aK)$. Similarly we get $T(K/a) \geq T(K)$. Thus, $T(aK)$ is equal to $T(K)$ for each $K \in \mathcal{K}$, so that $a = 1$.

Since

$$\Psi(a(s_n)K) = \Psi(K)/s_n \text{ and } \Psi(K/a(s_n)) = s_n\Psi(K),$$

it is sufficient to consider either case $a(s_n) \to 0$ or $a(s_n) \to \infty$ as $n \to \infty$.

Choose an integer $m > 1$. Let $a_m > 1$. Suppose that $a(s_n) \to \infty$ as $n \to \infty$. Then, for any $n \geq 1$,

$$a(s_n) = (a_m)^{k(n)}\Delta_n,$$

where $1 \leq \Delta_n < a_m$ and $k(n)$ is a certain positive integer. It follows from (1.6) and (1.9) that

$$(a_m)^{k(n)} = a_{m^{k(n)}}.$$

Hence

$$\begin{aligned} T(\Delta_n K) &= 1 - \left[1 - T\left((a_m)^{k(n)}\Delta_n K\right)\right] \\ &= 1 - \left[1 - T\left(a(s_n)K\right)\right]^{m^{k(n)}}. \end{aligned}$$

Let $0 < T(K)$. Then $T(\Delta_n K) \to 1$, since $T(a(s_n)K) \to T(K) > 0$. Without loss of generality suppose that $\Delta_n \to \Delta$ as $n \to \infty$. Semicontinuity of T implies $T(\Delta K) = 1$. Hence $\Delta K \cap F_A \neq \emptyset$ as soon as $T(K) > 0$. It is easy to show that $\Delta_n F_A = F_A$ for all $n \geq 1$, whence $\Delta F_A = F_A$. Thus, $K \cap F_A \neq \emptyset$ as soon as $T(K) > 0$, so that $A = F_A$ almost surely.

It is obvious that $a_m \neq 1$. If $a_m < 1$, then suppose $a(s_n) \to 0$ as $n \to \infty$ and use the same arguments as above.

Thus, $a(s_n) \to 1 = a(1)$ as $s_n \to 1$.

V. Let $s_n \to s \in \mathbb{Q}_+$ as $n \to \infty$. Then

$$a(s_n) = a(s)a(s_n/s) \to a(s) \text{ as } n \to \infty,$$

since $s_n/s \to 1$ and $a(s_n/s) \to 1$. Thus, the function $a(s)$ is continuous on $\mathbb{Q}_+$.

VI. Extend the function $a(s)$ onto the whole half-line. For any positive s denote $a(s) = \lim a(s_n)$, where $s_n \to s$ as $n \to \infty$, $s_n \in \mathbb{Q}_+$. Then the function $a(s)$ is continuous on $\mathbb{R}_+$, and $a(ss_1) = a(s)a(s_1)$ for each $s, s_1 > 0$. Thus, $a(s) = s^\gamma$ for a certain real γ. If $\gamma = 0$, then $s\Psi(K) = \Psi(K)$, i.e. $A = F_A$ almost surely. Hence $\gamma \neq 0$, i.e. (1.4) and (1.5) are valid for $\alpha = -1/\gamma$.

Sufficiency. The capacity functional of $A_1 \cup \cdots \cup_n$ is equal to

$$T_n(K) = 1 - \exp\{-n\Psi(K)\}.$$

On the other hand, the capacity functional of $a_n A$ is equal to

$$T'(K) = 1 - \exp\{-\Psi(K/a_n)\}.$$

If $a_n = n^{-1/\alpha}$, then $T_n = T'_n$ on $\mathcal{K}$. Thus, (1.2) follows from the Choquet theorem. □

Corollary 1.5 *A union-stable RACS A has no fixed points iff $\alpha > 0$ in (1.4). If $\alpha < 0$, then F_A is non-empty and $0 \in F_A$.*

PROOF. If $F_A = \emptyset$, then $\Psi(B_r(0)) < \infty$ for each $r > 0$. On the other hand, $\Psi(B_r(0)) = r^\alpha \Psi(B_1(0))$. Thus, $\alpha > 0$.

Let $\alpha > 0$. If F_A is non-empty, then, by (1.5), F_A contains the origin. On the other hand, $K \cap F_A = \emptyset$ yields $\Psi(sK) \to 0$ for $s \downarrow 0$, contrary to the semicontinuity of the capacity functional. □

Corollary 1.6 *A stationary U-stable RACS A has positive parameter α in (1.4).*

PROOF. Indeed, otherwise A has a fixed point, so that $A = \mathbb{R}^d$ by stationarity. □

Corollary 1.7 *For any $F \subset \mathbb{R}^d$ denote its inversion transformation*

$$F^* = \left\{u\|u\|^{-2}\colon\ u \in F\right\}. \tag{1.10}$$

Then a random closed set A is U-stable with parameter $\alpha \neq 0$ iff its inverse set A^* is U-stable with parameter $-\alpha$.

PROOF. Evidently,

$$\mathbf{P}\{A^* \cap K \neq \emptyset\} = \mathbf{P}\{A \cap K^* \neq \emptyset\} = 1 - \exp\{-\Psi(K^*)\}.$$

By (1.4), $\Psi(sK^*) = s^\alpha \Psi(K^*)$, whence $\Psi((sK)^*) = s^{-\alpha}\Psi(K^*)$. □

3.2 Examples of Union-stable Random Closed Sets.

It is easy to show that the random set $A = (-\infty, \xi] \subset \mathbb{R}^1$ is U-stable iff ξ is max-stable with parameter $\gamma \neq 0$, see (1.3). Consider other examples of U-stable sets related with Poisson point processes.

The *Poisson point process* Π_Λ with the intensity measure Λ is a random countable subset of $\mathbb{R}^d$ such that the following conditions are valid (see Stoyan et al., 1987).

1. For each bounded set Γ the random variable $\mathrm{card}(\Pi_\Lambda \cap \Gamma)$ has a Poisson distribution with parameter $\Lambda(\Gamma)$.
2. For all disjoint sets $\Gamma_1, \ldots, \Gamma_n$ the random variables $\mathrm{card}(\Pi_\Lambda \cap \Gamma_i)$ are independent, $1 \leq i \leq n$, for each $n \geq 2$.

Here Λ is a Borel measure on $\mathbb{R}^d$ called the *intensity measure.*

The capacity functional of Π_Λ is equal to

$$T(K) = \mathbf{P}\{\Pi_\Lambda \cap K \neq \emptyset\} = 1 - \exp\{-\Lambda(K)\}.$$

It is evident that any Borel measure satisfies the conditions **(T1)** and **(T2)** from Section 1.2. By Theorem 1.2, any Poisson point process is union-infinitely-divisible and the corresponding Choquet capacity $\Psi(K)$ coincides with the intensity measure $\Lambda(K)$. However, if Π_Λ is union-stable, then the measure Λ has to satisfy additional conditions.

EXAMPLE **2.1** Let A be the Poisson point process Π_Λ. Assume that Λ has the density λ with respect to the Lebesgue measure. Then A is U-stable iff λ is homogeneous, i.e.

$$\lambda(su) = s^{\alpha-d}\lambda(u) \tag{2.1}$$

for a certain real α, whatever u from $\mathbb{R}^d$ and positive s may be. If $\alpha < 0$, then the origin is a fixed point of A. For $\alpha = 0$ the random set A is the stationary Poisson point process.

The Poisson point process is of use to construct more complicated random closed sets called *Boolean* (or germ-grain) *models*, see Matheron (1975).

EXAMPLE **2.2** Let $\Pi_\Lambda = \{x_1, x_2, \ldots\}$ be points of the Poisson point process from Example 2.1, and let $A_0, A_0^1, A_0^2, \ldots$ be a sequence of independent identically distributed random sets in $\mathbb{R}^d$. Then the random closed set A defined as

$$A = \bigcup_{x_i \in \Pi_\Lambda} (x_i + A_0^i)$$

is said to be the *Boolean model* with the primary grain A_0, see Matheron (1975), Stoyan et al. (1987). It was shown in Matheron (1975) that the capacity functional of A is defined by (1.1), where

$$\Psi(K) = \mathbf{E}\Lambda(A_0 \oplus \check{K}),$$

$\check{K} = \{-x: x \in K\}$. If $sA_0 \stackrel{d}{\sim} A_0$ for each $s > 0$, then the Boolean model A is U-stable. We can choose A_0 to be a non-random cone, or the trajectory of a certain stable process, or the set $\{ex: x \in \mathfrak{Z}_0\}$, where e is a unit vector, $\mathfrak{Z}_0 = \{t \geq 0: w_t = 0\}$ is the zero set of the Wiener process. We only have to ensure the finiteness of $\mathrm{E}\Lambda(A_0 \oplus B_\varepsilon(x))$ for some $\varepsilon > 0$ and all x from $\mathbb{R}^d$. Thus, the parameter α in (1.4) and (2.1) is to be negative.

The following example is not related with Poisson point processes.

EXAMPLE 2.3 Let $f: \mathbb{R}^d \longrightarrow [0, \infty]$ be an upper semi-continuous function. Then

$$\Psi(K) = \sup_{x \in K} f(x) \tag{2.2}$$

is a maxitive Choquet capacity. The capacity functional (1.1) determines the distribution of the random closed set A defined as

$$A = \{x: f(x) \geq \xi\},$$

where ξ is a random variable, having the exponential distribution with parameter 1. The RACS A is U-stable iff the function f is homogeneous, i.e. $f(sx) = s^\alpha f(x)$ for each $s > 0$ and $x \in \mathbb{R}^d$. In this connection, $F_A = \{x: f(x) = \infty\}$.

Next, consider one quite general method of construction of capacities and, respectively, distributions of random sets. Let $k(x, y): \mathbb{R}^d \times \mathbb{R}^d \longrightarrow [0, \infty]$ be a lower semi-continuous function which is said to be a *kernel.* Furthermore, let

$$U_\mu^k(x) = \int_K k(x, y)\mu(dy)$$

denote the potential of the measure μ, and let S_μ be the support of μ. Assume that k satisfies the maximum principle, i.e. $U_\mu^k(x) \leq M$ for all $x \in S_\mu$ implies this inequality everywhere on $\mathbb{R}^d$. Then the functional

$$C(K) = \sup \left\{\mu(K): U_\mu^k(x) \leq 1, x \in K, S_\mu \subseteq K\right\}, K \in \mathcal{K}, \tag{2.3}$$

is a Choquet capacity on $\mathcal{K}$, see Landkof (1966) and Choquet (1953/54). Note that the supremum is taken over all measures satisfying the imposed conditions. For instance, the capacity Ψ defined in (2.2) can be obtained for $k(x, y) = 1/\max(f(x), f(y))$.

According to the Choquet theorem, the functional T defined as

$$T(K) = 1 - \exp\{-C(K)\} \tag{2.4}$$

is the capacity functional of a certain random closed set A. Then A is U-stable if and only if

$$k(sx, sy) = s^{-\alpha} k(x, y), \tag{2.5}$$

whatever $s > 0$ and $x, y \in \mathbb{R}^d$ may be. If A is stationary and isotropic, then $k(x, y)$ depends on $\|x - y\|$ only. Thus

$$k(x, y) = \begin{cases} C\|x - y\|^{-\alpha} & , \quad x \neq y \\ q & , \quad x = y \end{cases}$$

for a certain $\alpha \neq 0$.

Let us prove that $q = \infty$. If $q = 0$, then $C(\{x\}) = \infty$, so that $A = \mathbb{R}^d$ a.s. by stationarity. If $0 < q < \infty$, then $C(\{x\}) = 1/q$, $C(\{x,y\}) = 2(q + k(x,y))$. If $\alpha < 0$, then

$$\lim_{y \to x} C(\{x,y\}) = 2/q > C(x),$$

i.e. C is not upper semi-continuous. In case $\alpha > 0$ we get $C(\{x,y\}) < C(x)$ for sufficiently small $\|x-y\|$, i.e. C is not increasing. Thus, $q = \infty$ and

$$k(x,y) = C\|x-y\|^{-\alpha},\ x \neq y. \tag{2.6}$$

This kernel is equal up to a certain constant factor to the Riesz kernel $k_{d,\gamma}\|x-y\|^{\gamma-d}$ for $\gamma = d - \alpha$. Hence C is the Riesz capacity. It is known (see Landkof, 1966) that C is an alternating capacity of infinite order in case $0 < \alpha < 1$ for $d = 1$ or $d - 2 \leq \alpha < d$ for $d \geq 2$. Thus, the above described technique allows to construct examples of U-stable sets with such parameters. These U-stable sets can be described constructively as it follows, cf. Matheron (1975).

Let us construct the RACS A with the capacity functional (2.4) and kernel (2.6). Let μ be the equilibrium probability measure on $B_r(0)$ with respect to the kernel given by (2.6), i.e.

$$U_\mu^k(x) = 1/C(B_r(0)), x \in B_r(0),$$

and let N be the Poisson random variable of mean $C(B_r(0))$. At the moment $t = 0$ we launch N mutually independent and independent of N random stable processes $\xi_i(t)$, $1 \leq i \leq N$, with the index $d - \alpha$ and the initial distribution μ. Then A is the union of their trajectories. Indeed, the capacity functional of A is equal to

$$T(K) = 1 - \exp\{-C(B_r(0))T_1(K)\}.$$

Here $T_1(K)$ is the capacity functional of the random set A_1 defined to be the trajectory of one process $\xi_1(.)$. Let $\mathbf{P}_x$ be the distribution of the stable process which starts from x, $\mathfrak{m}_K = \inf\{t: \xi_1(t) \in K\}$. Since $\mathbf{P}_x\{\mathfrak{m}_K < \infty\}$ is the potential of the equilibrium measure μ_K on K (see Ito and McKean, 1965), we get

$$\begin{aligned} T_1(K) &= \int_{B_r} \mathbf{P}_x\{\mathfrak{m}_K < \infty\}\mu(dx) \\ &= \int_{B_r}\int_K k(x,y)\mu_K(dy)\mu(dx) \\ &= \int_K \mu_K(dy)/C(B_r(0)) \\ &= C(K)/C(B_r(0)). \end{aligned}$$

Hence A has the capacity functional (2.4).

It should be noted that the capacity functionals (2.4) for C given in (2.3) and the kernel (2.6) do not exhaust all examples of capacity functionals of U-stable stationary isotropic random closed sets.

3.3 Convex-Stable Random Sets.

The definition (1.2) of U-stable sets is rarely applicable to convex random sets since the set $A_1 \cup \cdots \cup A_n$ is usually non-convex. A convex RACS must satisfy very strong conditions to be U-stable.

Let A be a convex RACS in $\mathbb{R}^d$. This random set is said to be *C-stable (convex-stable)* if, for every $n \geq 2$ and independent copies $A_1, \ldots, A_n$ of A,

$$a_n A \overset{d}{\sim} \operatorname{conv}(A_1 \cup \cdots \cup A_n) \oplus K_n \tag{3.1}$$

for some $a_n > 0$, $K_n \in \mathcal{K}$. In case $K_n = \{b_n\}$, $n \geq 1$, i.e. K_n is a singleton set for all $n \geq 1$, the random set A is said to be *strictly C-stable.*

This definition is due to Gine, Hahn and Vatan (1990), where such a set A is additionally supposed to be compact and its support function is assumed to have a non-degenerate distribution.

Convex-stable random sets arise in applied sciences while treating objects determined by convex hulls of their elementary components. For example, a star-cluster can be considered to be the convex hull of the stars or sub-clusters, the natural habitat of a certain species is the convex hull of the sightings of animals, the dangerous region to be placed in quarantine in epidemiology is the convex hull of the primary regions where the infectious disease has manifested. The corresponding random sets are C-stable, since they arise as limits for convex hulls (see Chapter 8).

It should be noted that convex hulls of random samples were studied in the theory of approximations of convex sets (see Schneider, 1988) and in statistics while testing for lack of circular symmetry (see Davis et al., 1988).

Reformulating (3.1) in terms of support functions, we obtain

$$a_n s_A(u) \overset{d}{\sim} \max\{s_{A_1}(u), \ldots, s_{A_n}(u)\} + s_{K_n}(u), \; u \in \mathbb{S}^{d-1}. \tag{3.2}$$

If A is *compact* and $s_A(u)$ has *non-degenerate* distribution for each $u \in \mathbb{S}^{d-1}$, then the support function $s_A(\cdot)$ is a random max-stable sample continuous process. This fact leads to the characterization of compact C-stable sets, see Gine et al. (1990). Meanwhile, even for $\mathbb{R}^1$ these conditions are more restrictive than it seems to be.

Example 3.1 Let ξ_1, ξ_2, ξ_3 be max-stable random variables with distributions of types I, II, III respectively, see (1.3) and Galambos (1978). Then the random sets $A_1 = [0, \xi_1]$, $A_2 = (-\infty, \xi_2]$, $A_3 = (-\infty, \xi_3]$ are even strictly C-stable. However these sets do not fall in with the scheme given in Gine et al. (1990), since A_2 and A_3 are unbounded and the support function $s_{A_1}(u)$ is degenerated for $u = -1$. Respectively, these sets cannot be represented as the sum of a certain non-random compact and a C-stable set whose interior almost surely contains the origin (this representation is an essential result of Gine et al., 1990).

Below we characterize C-stability of random set in terms of its inclusion functional ι. We show also that slightly modified methods proposed by Gine et al. (1990) work well even in general case.

First, give a modified version of Lemma 3.3 from Gine et al. (1990).

Theorem 3.2 *Let Ξ be a random closed subset of a σ-compact space satisfying*

$$\Xi \overset{d}{\sim} \Xi_1 \cap \cdots \cap \Xi_n, n \geq 1,$$

for Ξ_i iid copies of Ξ. Then Ξ almost surely coincides with the set of its fixed points.

For any RACS A define

$$\begin{aligned} L_A &= \left\{u \in \mathbb{S}^{d-1}\colon \mathbf{P}\left\{s_A(u) < \infty\right\} = 1\right\}, \\ H_A &= \left\{\alpha u + \beta v\colon\ u, v \in L_A, \alpha, \beta \in \mathbb{R}\right\}, \end{aligned}$$

i.e. H_A is the minimal linear span of L_A.

Theorem 3.3 *If A is a C-stable RACS, then*

$$A = A_0 \oplus H_A^{\perp} \tag{3.3}$$

where $A_0 = A \cap H_A$ is a C-stable subset of H_A, $H_A^{\perp}$ is the orthogonal complement to H_A.

PROOF. Put

$$\Xi_A = \left\{u \in \mathbb{S}^{d-1}\colon\ s_A(u) < \infty\right\}.$$

It can be shown that Ξ_A is a random closed subset of $\mathbb{S}^{d-1}$, and

$$\Xi_A \overset{d}{\sim} \Xi_{\mathrm{conv}(A_1,\ldots,A_n)\oplus K_n} = \bigcap_{i=1}^{n} \Xi_{A_i}.$$

Theorem 3.2 yields $\Xi_A = L_A$ a.s., since L_A is the set of fixed points of Ξ_A. Furthermore, $A_0 \neq \emptyset$ a.s., since otherwise $s_A(u)$ is finite with positive probability for a certain u from $H_A^{\perp}$. Hence, (3.3) is valid. □

Corollary 3.4 *The distribution of a C-stable RACS is determined uniquely by the inclusion functional $\mathfrak{t}(F)$, $F \in \mathcal{C}$.*

PROOF is obvious, since the functional $\mathfrak{t}(F)$ determines uniquely finite-dimensional distributions of $s_A(u)$, $u \in L_A$, and, therefore, the distribution of A, cf. Example 1.3.2. □

The following theorem provides a characterization of inclusion functionals of C-stable sets. The class $\mathcal{C}_A$ defined as

$$\mathcal{C}_A = \left\{F \oplus H_A^{\perp}\colon\ F \in \mathcal{C}, F_A \cap H_A \subseteq F \subseteq H_A\right\}$$

plays the same role as the class $\mathcal{K}_A$ in Theorem 1.4. It follows from (3.3) that $\mathfrak{t}(F) = 0$ for F belonging to $\mathcal{C} \setminus \mathcal{C}_A$.

Theorem 3.5 *A non-trivial convex RACS A is $\mathcal{C}$-stable if and only if its inclusion functional $\mathfrak{t}$ is equal to*

$$\mathfrak{t}(F) = \exp\{\psi(F)\}, F \in \mathcal{C}, \tag{3.4}$$

where ψ is a non-positive functional satisfying conditions **(I1)**, **(I2)** *(see Section 1.3) and $\psi(F) > -\infty$ iff $F \in \mathcal{C}_A$. In addition, for a certain convex compact $H \subset H_A$ and $\gamma \neq 0$, one of the following two groups of relations is valid*

$$\psi(F) = s\psi(F \ominus H \log s), \; F_A \ominus H \log s = F_A, \tag{3.5}$$

$$\psi(F) = s\psi(s^\gamma F \ominus (s^\gamma - 1)H), \; (F_A \ominus H)s = F_A, \tag{3.6}$$

whatever positive s and F from $\mathcal{C}_A$ may be. If L_A consists of at least two centrally symmetric points, then only (3.6) is valid with $\gamma > 0$.

PROOF. *Necessity.* It follows from Theorem 4.2 of Gine et al. (1990) that there exists a measure ν on $\mathcal{C}$ such that

$$\nu(\{M \in \mathcal{C}\colon\ M \not\subset F\}) = -\log \mathfrak{t}(F).$$

Then it is easy to show that the functional

$$\psi(F) = -\nu(\{M \in \mathcal{C}\colon\ M \not\subset F\})$$

satisfies conditions **(I1)**, **(I2)** and is finite on $\mathcal{C}_A$, see also Trader (1981). In addition, we have to derive (3.5) and (3.6) from (3.1).

The random process $s_A(u)$, $u \in L_A$, is continuous max-stable on L_A. It follows from (3.2) and results of Gine et al. (1990) that $a_n = n^\gamma$ for some real γ. If $\gamma = 0$, then

$$s_{K_n}(u) = -h(u) \log n, u \in L_A.$$

Suppose that L_A does not consists of any pair of centrally symmetric points. Then there exists a compact set H such that $K_n = H \log n$, $n \geq 1$ (the set H, e.g., may be chosen to be centrally symmetric). Hence

$$A \overset{d}{\sim} \operatorname{conv}(A_1 \cup \cdots \cup A_n) \oplus H \log n, \tag{3.7}$$

and

$$n\psi(F \ominus H \log n) = \psi(F)$$

for all $n \geq 1$ and $F \in \mathcal{C}_A$. Thus, for any $q = m/n > 0$

$$n\psi(F \ominus H \log n) = m\psi(F \ominus H \log m).$$

Put $F = F' \oplus H \log n$ for a certain F' from $\mathcal{C}_A$. Then

$$q\psi(F' \ominus H \log q) = \psi(F'), G \in \mathcal{C}_A.$$

Here the Minkowski subtraction is replaced with the addition for $q < 1$. By semi-continuity, (3.5) is valid for any positive q. It follows from (3.7) that $F_A \oplus H \log n = F_A$, so that

$$F_A \oplus H \log q = F_A, \; q > 0.$$

If $\gamma \neq 0$, then

$$s_{K_n}(u) = (n^\gamma - 1)h(u),\ u \in L_A,$$

so that $K_n = (n^\gamma - 1)H$. Similar arguments as above lead to (3.6). If L_A consists of two points $\{y, -y\}$ then the proof of Proposition 4.4 from Gine et al. (1990) is applicable, so that the possibility of $\gamma \leq 0$ is rejected.

Sufficiency immediately follows from Corollary 3.4. □

The characterization theorem for compact C-stable sets follows from Theorem 3.3, since in this case $L_A = \mathbb{S}^{d-1}$ and, evidently, L does contain centrally symmetric points. This case ($\gamma > 0$) was considered in Gine et al. (1990). Besides, a spectral representation for the inclusion functional $\mathfrak{t}$ was obtained. It was proven that each C-stable set with $\gamma > 0$ and nondegenerately distributed support function can be represented as $\Theta \oplus H$, where Θ is a C-stable random set containing the origin as an interior point almost surely.

We shall say that the C-stable RACS A belongs to the first type if (3.5) is valid and to the second type otherwise. If A belongs to the second type, then (3.6) is transformed to

$$\psi(sF \oplus H) = s^\alpha \psi(F \oplus H), s > 0, F \in \mathcal{C}_A, \tag{3.8}$$

where $\alpha = -1/\gamma$.

It should be noted that $H \supset F_A$ in case $\gamma > 0$, and H is a singleton if A is strictly C-stable.

Corollary 3.6 *Let A be the C-stable set of the second type. If $0 < \mathfrak{t}(F \oplus H) < 1$ for some convex F containing the origin, then the parameter α in (3.8) is negative. If $0 < \mathfrak{t}(F \oplus H) < 1$ for some $F \in \mathcal{C}$ such that $0 \notin F$, then $\alpha > 0$.*

PROOF. If $0 < \mathfrak{t}(F \oplus H) < 1$ and $0 \in F$, then

$$s_2 F \supset s_1 F,\ s_2 \geq s_1 > 0.$$

Hence $s^\alpha \psi(F \oplus H)$ increases and $-\infty < \psi(F \oplus H) < 0$. Therefore, $\alpha < 0$.

Note that $\mathfrak{t}(F_0 \oplus H) < 1$ for any half-space F_0 missing the origin. In fact, $\mathfrak{t}(F_0 \oplus H) = 1$ implies $\mathfrak{t}(sF_0 \oplus H) = 1$ for sufficiently large s, so that $A = \emptyset$. Suppose that $0 < \mathfrak{t}(F) < 1$, and $0 \notin F$. Then $\mathfrak{t}(F_0 \oplus H) > 0$ for a certain half-space F_0 such that $0 \notin F_0$, $F \subset F_0$. Thus, $0 < \mathfrak{t}(F_0 \oplus H) < 1$ and $s_2 F_0 \subset s_1 F_0$ for $s_2 \geq s_1 > 0$. Hence $s^\alpha \psi(F_0 \oplus H)$ decreases for $s > 0$, and also $-\infty < \psi(F_0 \oplus H) < 0$. Therefore, α is positive. □

EXAMPLE **3.7** Let $\xi = (\xi_1, \ldots, \xi_d)$ be a random vector in $\mathbb{R}^d$. Put

$$A = \operatorname{conv}\{e_i x_i \colon 1 \leq i \leq d, x_i \leq \xi_i\},$$

where $e_1, \ldots, e_d$ is the basis in $\mathbb{R}^d$. Then A is strictly C-stable iff ξ is max-stable random vector with respect to coordinate-wise maximum, see Galambos (1978), Balkema and Resnick (1977). Evidently,

$$\mathfrak{t}(C(x)) = \mathbf{P}\{A \subseteq C(x)\} = F_\xi(x),$$

where F_ξ is the distribution function of ξ,

$$C(x) = (-\infty, x_1] \times \cdots \times (-\infty, x_d],\ x = (x_1, ..., x_d).$$

Thus, A is strictly C-stable iff $F_\xi(x) = \exp\{\psi(x)\}$, where $\psi(x) \leq \psi(y) \leq 0$ if $x \leq y$ coordinate-wisely, and one of the following conditions is valid for a certain $v \in \mathbb{R}^d$

$$\begin{aligned}\psi(x) &= s\psi(x + v\log s),\\ \psi(x) &= s\psi(s^\gamma x + (s^\gamma - 1)v),\end{aligned}$$

whatever positive s and $x \in \mathbb{R}^d$ may be.

EXAMPLE 3.8 Let η be the exponential random variable with parameter λ, $f: \mathbb{R}^d \longrightarrow [0, +\infty]$ be a certain upper semi-continuous function. Then

$$A = \operatorname{conv}\left\{x \in \mathbb{R}^d:\ f(x) \geq \eta\right\}$$

is a convex RACS. Its inclusion functional is equal to

$$\mathfrak{i}(F) = \mathbf{P}\left\{\sup_{x \in F^c} f(x) < \eta\right\} = \exp\{\psi(F)\}$$

for $\psi(F) = -\lambda \sup\{f(x): x \in F^c\}$. Note that $F_A = \operatorname{conv}\{x \in \mathbb{R}^d: f(x) = +\infty\}$. The set A is of the first type iff $sf(x + v\log s) = f(x)$, and of the second type iff $sf(s^\gamma x + (s^\gamma - 1)v) = f(x)$ for certain $v \in \mathbb{R}^d$, $\gamma \neq 0$, whatever $x \in \mathbb{R}^d$ and $s > 0$ may be.

It should be noted that for any U-stable RACS A its convex hull $\operatorname{conv}(A)$ (if non-trivial) is strictly C-stable.

EXAMPLE 3.9 Let Π_Λ be a Poisson point process from Example 2.1, with $\alpha < 0$. Then the random set $A = \operatorname{conv}(\Pi_\Lambda)$ is C-stable. Its inclusion functional is evaluated as

$$\mathbf{P}\{A \subseteq F\} = \exp\{-\Lambda(F^c)\}, F \in \mathcal{C}.$$

3.4 Generalizations and Remarks.

A possible generalization of the notion of U-stable random sets is based on the analog of (3.1). A random set A is said to be *generalized union-stable* (GU-stable) if, for any collection $A_1, ..., A_n$ of independent copies of A,

$$a_n A \overset{d}{\sim} (A_1 \cup \cdots \cup A_n) \oplus K_n \tag{4.1}$$

for certain $a_n > 0$, $K_n \in \mathcal{K}$. A random set is said to be strictly GU-stable if it satisfies (4.1) with single-point compacts $K_n = \{b_n\}$.

Generalized union-stable random sets are very difficult to characterize, since, in general, the characterization problem cannot be reduced to examination of max-stable support functions. The situation is getting worse in case A is unbounded, since in this case A may coincide in distribution with $A + u$ for some $u \neq 0$. The main obstacle is the lack of the Khinchin lemma (see Leadbetter et al., 1986) for random closed sets. Namely,

$$T(a_nK + b_n) \to T(K) \text{ as } n \to \infty$$

for each K from $\mathcal{K}$ does not imply the boundedness of a_n, $\|b_n\|$, $n \geq 1$.

It was proven in Trader (1981) that any strictly GU-stable random set A with $a_2 \neq 1$ can be reduced to a U-stable random set by means of a non-random shift. In other words, there exists $b \in \mathbb{R}^d$ such that $A + b$ is U-stable. It was proven also that in this case a_n is equal to n^γ, $n \geq 1$.

Consider the special case of (4.1) for $a_n = 1$ and $K_n = \{-b_n\}$. The random set A is said to be *additive union-stable* (AU-stable) if

$$A + b_n \overset{d}{\sim} (A_1 \cup \cdots \cup A_n). \tag{4.2}$$

Let

$$H_A = \left\{u \in \mathbb{R}^d \colon A \overset{d}{\sim} A + u\right\} \tag{4.3}$$

be the set of all invariant shifts for A. Suppose that H_A is a cone and A is not stationary, i.e. $H_A \neq \mathbb{R}^d$.

The RACS A is said to be *homogeneous at infinity* if, for each sequence $\{b_n, n \geq 1\} \subset \mathbb{R}^d$,

$$\lim_{n\to\infty} T(K + b_n) = \lim_{n\to\infty} T(K + u + b_n), K \in \mathcal{K}, u \in \mathbb{R}^d, \tag{4.4}$$

as soon as at least one limit exists, and $\rho(b_n, H_A) \to \infty$ as $n \to \infty$. It is evident that any compact random set is homogeneous at infinity.

Lemma 4.1 *If A is homogeneous at infinity, then A satisfies the Khinchin lemma, i.e. for each sequence $b_n \in \mathbb{R}^d$, $n \geq 1$, the pointwise convergence $T(b_n + K) \to T(K)$ as $n \to \infty$ for $K \in \mathcal{K}$ implies* $\sup\{\|b_n\| \colon n \geq 1\} < \infty$.

PROOF. Suppose that $\rho(b_n, H_A) \to \infty$ as $n \to \infty$. It follows from (4.4) that the limit of $T(K + u + b_n)$ exists and is equal to $T(K)$. On the other hand $T(K + u + b_n) \to T(K + u)$ as $n \to \infty$. Thus $T(K) = T(K + u)$ for each u, i.e. $H_A = \mathbb{R}^d$. Hence the sequence b_n, $n \geq 1$, is bounded. □

Theorem 4.2 *The RACS A is AU-stable if (and only if in case A is homogeneous at infinity) its capacity functional is of the form (1.1), where $\Psi(\emptyset) = 0$, and, for a certain $v \perp H_A$,*

$$\Psi(K + vs) = e^{-s}\Psi(K), \quad F_A + vs = F_A, \tag{4.5}$$

whatever K from $\mathcal{K}_A$ and s from $\mathbb{R}$ may be.

PROOF. *Sufficiency* can be obtained from (4.5) and (4.1) for $b_n = v \log n$.

Necessity. Since A is union-infinitely-divisible, (1.1) is valid. It is easy to prove the existence of an $H_A^{\perp}$-valued function $b(s)$, $s \in \mathbb{Q}_+$, such that

$$s\Psi(b(s) + K) = \Psi(K), \ K \in \mathcal{K}_A, \ b(s) + F_A = F_A. \tag{4.6}$$

As in the proof of Theorem 1.4, we can show that

$$b(ss_1) = b(s) + b(s_1)$$

for all positive rational numbers s, s_1. It follows from (4.6) that $T(b(s_n)+K) \to T(K)$, $K \in \mathcal{K}$, as soon as $s_n \to 1$, $n \to \infty$. Lemma 4.1 yields

$$\sup\{\|b(s_n)\|: n \geq 1\} < \infty.$$

Without loss of generality assume that $b(s_n) \to b$ as $n \to \infty$. It is easy to show that $T(b+K) = T(K)$ for each $K \in \mathcal{K}$, and $b = 0$. Thus, $b(s)$ is continuous at $s = 1$ and, therefore, may be continuously extended onto the positive half-line. Hence $b(s) = v \log s$ for a certain vector $v \perp H_A$, so (4.5) follows from (4.6). □

Example 4.3 Let $A = (-\infty, \xi]$. Then A is AU-stable iff ξ is a max-stable random variable with distribution (1.3), $\gamma = 0$.

Note that the random closed sets from Examples 2.1 and 2.2 are AU-stable as long as

$$\lambda(u + vs) = e^{-s}\lambda(u), u \in \mathbb{R}^d, s \in \mathbb{R}$$

for a certain vector v belonging to $\mathbb{R}^d \setminus \{0\}$.

Mention several open problems related with characterization of generalized union-stable sets.

1. Characterization of generalized union-stable sets for one-point compacts K_n can be obtained in case the random sets are supposed to satisfy an analog of the Khinchin lemma. It can be shown that either $A + v$ is union-stable for a certain vector v, or A is additive union-stable. The problem is to find examples of GU-stable sets which do not satisfy the Khinchin lemma.

2. It seems interesting to characterize GU-stable sets for arbitrary compacts K_n and, in particular, for circular K_n.

3. The random set A is said to be inverted union-stable if

$$a_n A \oplus K_n \overset{d}{\sim} A_1 \cup \cdots \cup A_n \tag{4.7}$$

in the above introduced notations. The matter is that an inverted union-stable set is no longer infinitely divisible by definition, so that we cannot use the representation theorem for capacities of infinitely-divisible random sets. The conjecture is that in this case $(A \ominus K_n) \oplus K_n$ converges to A a.s. in the Hausdorff metric as $n \to \infty$, so that A can be approximated by union-infinitely-divisible random sets. The problem is to find a representation theorem for their capacity functionals and to characterize random closed sets which do not satisfy (4.1), but (4.7). The same problem arises for inverted convex stable sets.

4. The statement of Lemma 4.1 is valid for homogeneous at infinity random sets. The problem is to find other conditions, which implies this statement (i.e. to prove the Khinchin lemma for random sets).

5. Find out examples of non-strictly C-stable random sets defined by random vectors (see Example 3.7). Which vectors are stable with respect to this definition?

Chapter 4

Limit Theorems for Normalized Unions of Random Closed Sets

4.1 Sufficient Conditions for the Weak Convergence of Unions of Random Sets.

In this chapter we consider limit theorems for *normalized unions of random sets*, where U-stable sets appear as limits. The reader is referred to Section 1.4 for generalities on the weak convergence of random closed sets.

Let $A_1, \ldots, A_n$ be independent identically distributed random closed sets with the common capacity functional T, and let X be their union, i.e.

$$X_n = A_1 \cup \ldots \cup A_n.$$

We investigate the weak convergence of $a_n^{-1} X_n$ where a_n, $n > 0$, is a suitable sequence of real numbers. It is evident that the limiting RACS X (if exists) is U-stable. Hence its capacity functional $\tilde{T}$ is characterized by Theorem 3.1.4, i.e.

$$\tilde{T}(K) = 1 - \exp\{-\Psi(K)\} \tag{1.1}$$

for a certain *homogeneous* capacity Ψ such that $\Psi(sK) = s^\alpha \Psi(K)$, $s > 0$, for each compact K missing the set of fixed points F_X. We shall show that the corresponding parameter α is positive in case $a_n \to 0$ and is negative if $a_n \to \infty$.

The *union-scheme* generalizes well-known limit theorems for normalized extremes of random variables. For instance, if $A_i = (-\infty, \xi_i]$, then $a_n^{-1} X_n$ converges weakly as soon as the random variable $a_n^{-1} \max(\xi_1, \ldots, \xi_n)$ has a weak limit. Thus, the limit theorems for scaled extremes of random variables will follow from our results for $A = (-\infty, \xi]$ or $A = [\xi, +\infty)$. Naturally, while handling with unions of random sets we use similar methods as in the theory of extremes, see e.g. Galambos (1978). However, the direct generalization fails due to specific properties of capacities. For example, the function $1 - T(xK)$, $x > 0$, plays in our consideration the same role as the distribution function in limit theorems for extremes, but this function is no longer monotone and even may not tend to 1 as $x \to \infty$.

First, consider the case $a_n \to \infty$ as $n \to \infty$. The limiting random set X has the origin as a fixed point, so that Corollary 3.1.5 yields $\alpha < 0$. For any compact K put

$$a_n(K) = \sup\{x\colon\ T(xK) \geq 1/n\}, \tag{1.2}$$

where $a_n(K) = 0$ in case $T(xK) < 1/n$ for all $x > 0$. The values $a_n(K)$, $n \geq 1$, stand for the "best" norming constants for the given compact K, see Molchanov (1993e).

The reader is referred to Section 1.6 for necessary facts from regular variation theory.

Introduce the function $\tau_K(x)$, $x \geq 0$, and the class $\mathcal{T}$ of compacts by

$$\tau_K(x) = T(xK), \tag{1.3}$$

$$\mathcal{T} = \left\{K\colon \liminf_{x\to\infty} T(xK) = 0\right\}. \tag{1.4}$$

Theorem 1.1 *Assume that for any K from $\mathcal{T}$ there exists the limit of $a_n(K)/a_n$ (which is not necessary finite), and let $\tau_K(x)$ be a regularly varying function with the negative exponent α. Then $a_n^{-1}X_n$ converges weakly to the U-stable set X with the capacity functional given by (1.1) for*

$$\Psi(K) = \begin{cases} \lim(a_n(K)/a_n)^{-\alpha} & , \quad K \in \mathcal{T} \\ \infty & , \quad \textit{otherwise} \end{cases} \tag{1.5}$$

We begin with a lemma.

Lemma 1.2 *Let $f(x)$ be a regularly varying function,* ind$f$ $= \alpha < 0$, *and let $g(x)$ be a non-negative function such that $xg(x) \to \infty$ as $x \to \infty$ and $g(x)$ has a certain (maybe infinite) limit as $x \to \infty$. Then*

$$\lim_{x\to\infty} \frac{f(xg(x))}{f(x)} = \lim_{x\to\infty} g(x)^\alpha. \tag{1.6}$$

PROOF follows from the representation of a regularly varying function, see Seneta (1976) and Section 1.6. Namely, $f(x) = x^\alpha L(x)$, where $L(x)$ is a slowly varying function, and also for a certain $B > 0$:

$$L(x) = \exp\left\{\eta(x) + \int_B^x \frac{\varepsilon(t)}{t} dt\right\}, x \geq B.$$

Here $\varepsilon(t)$ tends to zero as $t \to \infty$, and $\eta(x)$ has the finite limit as $x \to \infty$. If $g(x)$ possesses a finite positive limit, then (1.6) is obvious, see also Theorem 1.1 from Seneta (1976).

Suppose that $g(x) \to \infty$ as $x \to \infty$. Let $c \in (0, -\alpha)$ be specified. Then $g(x)^\alpha \to 0$ as $x \to \infty$. Hence

$$\begin{aligned}
\lim_{x\to\infty} \frac{f(xg(x))}{f(x)} &= \lim_{x\to\infty} g(x)^\alpha \exp\left\{\eta(g(x)x) - \eta(x) + \int_B^{g(x)x} \frac{\varepsilon(t)}{t} dt\right\} \\
&= \lim_{x\to\infty} g(x)^\alpha \exp\left\{\int_B^{g(x)x} \frac{\varepsilon(t)}{t} dt\right\} \\
&\leq \lim_{x\to\infty} g(x)^\alpha \exp\{c \log g(x)\} \\
&= \lim_{x\to\infty} g(x)^{\alpha+c} = 0.
\end{aligned}$$

Let $g(x) \to 0$ as $x \to \infty$. Then $g(x)x \geq B$ for sufficiently large x, and

$$\begin{aligned}\lim_{x\to\infty}\frac{f(xg(x))}{f(x)} &= \lim_{x\to\infty} g(x)^{\alpha}\exp\left\{-\int_{g(x)x}^{x}\frac{\varepsilon(t)}{t}dt\right\} \\ &\geq \lim_{x\to\infty} g(x)^{\alpha}\exp\left\{-\int_{g(x)x}^{x}\frac{c}{t}dt\right\} \\ &= \lim_{x\to\infty} g(x)^{\alpha+c} = \infty.\end{aligned}$$

Thus, (1.6) is valid too. □

PROOF OF THEOREM 1.1. If T is the capacity functional of the random set A_1, then the RACS $a_n^{-1}X_n$ has the capacity functional T_n given by

$$T_n(K) = 1 - (1 - T(a_nK))^n. \tag{1.7}$$

If $K \notin \mathcal{T}$, then $T_n(K) \to 1 = \tilde{T}(K)$ as $n \to \infty$, i.e. (1.5) is valid. Further suppose that $K \in \mathcal{T}$. It follows from (1.7) that

$$\tilde{T}(K) = \lim_{n\to\infty} T_n(K) = 1 - \exp\{-\Psi(K)\}, \tag{1.8}$$

as long as the limit

$$\lim_{n\to\infty} nT(a_nK) = \Psi(K) \tag{1.9}$$

exists (it may be infinite).

Suppose that

$$\limsup_{x\to\infty} T(xK) \geq \varepsilon > 0,$$

cf. (1.4). Then $a_n(K) = \infty$ for all $n \geq n_0$, i.e.

$$q_n(K) = a_n(K)/a_n = \infty,\ n \geq n_0.$$

Let $\lambda > 1$ be specified. Then for each $n \geq n_0$ there exists $\lambda_n > \lambda$ such that $T(a_n\lambda_nK) \geq 1/n$. Hence

$$\begin{aligned}\lim_{n\to\infty} nT(a_nK) &= \lim_{n\to\infty} n\frac{T(a_n\lambda_nK)\tau_K(a_n)}{\tau_K(a_n\lambda_n)} \\ &\geq \liminf_{n\to\infty}\lambda_n^{-\alpha} \\ &\geq \lambda^{-\alpha}.\end{aligned}$$

Letting λ go to infinity yields

$$\lim_{n\to\infty} nT(a_nK) = \infty.$$

Hence $\tilde{T}(K) = 1$, i.e. (1.5) is valid with $\Psi(K) = \lim q_n(K) = \infty$.

Let $\tau_K(x) \to 0$ as $x \to \infty$. In this case $a_n(K) < \infty$ for all $n \geq 1$. If the sequence $a_n(K)$, $n \geq 1$, is bounded, then $T(xK) = 0$ for all sufficiently large x, so that $\lim T_n(K) = 0$. Hence (1.5) is valid, since

$$\Psi(K) = \lim_{n\to\infty}(a_n(K)/a_n)^{-\alpha} = 0.$$

Further assume that $a_n(K) \to \infty$ as $n \to \infty$. Lemma 1.2 yields

$$\lim_{n\to\infty} \frac{\tau_K(a_n)}{\tau_K(a_n(K))} = \lim_{n\to\infty} \frac{\tau_K(a_n)}{\tau_K(q_n(K)a_n)} = \lim_{n\to\infty} (q_n(K))^{-\alpha}. \tag{1.10}$$

Let us prove that $nT(a_n(K)K) \to 1$ as $n \to \infty$. For arbitrary $n \geq 1$ choose x_m, $m \geq 1$, such that

$$a_n(K) - 1/m \leq x_m \leq a_n(K)$$

and $T(x_m K) \geq 1/n$. Semi-continuity of T implies

$$1/n \leq \lim_{m\to\infty} T(x_m K) \leq T(a_n(K)K).$$

Thus, $nT(a_n(K)K) \geq 1$ for all $n \geq 1$. It follows from (1.2) that $T(a_n(K)\lambda K) \leq 1/n$ for $\lambda > 1$. Regular variation of τ_K implies

$$1 \leq \lim_{n\to\infty} nT(a_n(K)K) = \lim_{n\to\infty} nT(a_n(K)\lambda K) \frac{\tau_K(a_n(K))}{\tau_K(\lambda a_n(K))} \leq \lambda^{-\alpha}.$$

Letting λ go to 1 yields

$$\lim_{n\to\infty} nT(a_n(K)K) = 1.$$

From (1.10) we get

$$\begin{aligned} \lim_{n\to\infty} nT(a_n(K)K) &= \lim_{n\to\infty} nT(a_n(K)K) \frac{\tau_K(a_n)}{\tau_K(a_n(K))} \\ &= \lim_{n\to\infty} \left(\frac{a_n(K)}{a_n}\right)^{-\alpha}. \end{aligned}$$

Thus, (1.5) is valid for every compact K. It is easy to verify that for any $K \in \mathcal{T}$ and $s > 0$ the compact sK belongs to $\mathcal{T}$, and also

$$\Psi(sK) = \lim_{n\to\infty} \left(\frac{a_n(sK)}{a_n}\right)^{-\alpha} = s^\alpha \Psi(K).$$

Therefore, the limiting random set is U-stable with parameter α. □

The limiting random set X in the scheme of Theorem 1.1. has the origin as a fixed point, since $a_n(B_r(0)) = \infty$ for each $r > 0$.

If the norming factor a_n converges to a certain positive constant a, then $a_n^{-1}X_n$ converges almost surely in the Hausdorff metric to the non-random set M defined as is the closure of

$$\bigcup \left(\{aB_r(x) \colon T(B_r(x)) = 0, r > 0, x \in \mathbb{R}^d\} \right).$$

It was stated in Section 1.4 that the weak convergence of random closed sets follows from the pointwise convergence of their capacity functionals on the subclass $\mathcal{K}_{ub} \subseteq \mathcal{K}$. This class $\mathcal{K}_{ub}$ consists of all finite unions of balls with positive radii. Hence $a_n^{-1}X_n$ converges weakly even in case the conditions of Theorem 1.1 are valid on $\mathcal{T} \cap \mathcal{K}_{ub}$ instead of $\mathcal{T}$.

Below we provide only the outline of the limit theorem in case $a_n \to 0$ as $n \to \infty$. The limiting random set X is U-stable with parameter $\alpha > 0$.

Denote for $K \in \mathcal{K}$

$$\bar{a}_n(K) = \inf\{x \geq 0\colon T(xK) \geq 1/n\}, \tag{1.11}$$

where $\bar{a}_n(K) = \infty$ in case $T(xK) < 1/n$ for all $x > 0$. Let

$$\bar{\mathcal{T}} = \left\{K\colon \liminf_{x\to 0} T(xK) = 0\right\}.$$

The function $f(x)$ is said to be *regularly varying at zero* with exponent α if $\tilde{f}(x) = f(1/x)$ is a regularly varying function with $\operatorname{ind}\tilde{f} = -\alpha$.

Theorem 1.3 *Assume that for any K from $\bar{\mathcal{T}}$ there exists the limit of $\bar{a}_n(K)/a_n$ (which is not necessary finite) and let the function $\tau_K(x)$ be regularly varying at zero with exponent $\alpha > 0$. Then $a_n^{-1}X_n$ converges weakly to the U-stable set X with capacity functional (1.1) where*

$$\Psi(K) = \begin{cases} \lim(\bar{a}_n(K)/a_n)^{-\alpha} & , \quad K \in \bar{\mathcal{T}}. \\ \infty & , \quad \text{otherwise} \end{cases}. \tag{1.12}$$

PROOF is quite similar to the proof of Theorem 1.1. □

4.2 Necessary Conditions in Limit Theorems for Unions.

It is well-known that regular variation conditions are both sufficient and necessary in limit theorems for extremes of random *variables*, see Galambos (1978). However, for random *sets* the situation is different to some extent, since the pointwise convergence of $T_n(xK)$ for all positive x no longer implies the uniform convergence. Nevertheless, the sufficient conditions in the scheme of Theorems 1.1, 1.3 are very close to the necessary ones.

Theorem 2.1 *Let the capacity functional T_n of $a_n^{-1}X$ converge uniformly on $\mathcal{K}$ to the capacity functional $\tilde{T}$ of a U-stable random closed set X with parameter α, see (1.1). Consider a compact K such that $\tilde{T}(K) < 1$ (i.e. K misses the set of fixed points of X). If $\alpha < 0$, then $K \in \mathcal{T}$ and the function $\tau_K(x)$ is regularly varying with exponent α. If $\alpha > 0$ then $K \in \bar{\mathcal{T}}$, and $\tau_K(x)$ is regularly varying at zero with the same exponent α.*

PROOF. Let T be the capacity functional of A_1. Denote

$$F(x) = \begin{cases} 1 - T(xK) & , \quad x > 0 \\ 0 & , \quad \text{otherwise} \end{cases}.$$

Since $\tilde{T}(K) < 1$, the corresponding functional $\Psi(K)$ from (1.1) is finite. It follows from (1.7) that

$$F^n(a_n x) \to \exp\{-\Psi(xK)\} = \exp\{-x^\alpha \Psi(K)\} = \tilde{F}(x) \text{ as } n \to \infty \tag{2.1}$$

uniformly for $x > 0$. If $\alpha < 0$, then $\tilde{F}(x)$ is the distribution function of a certain max-stable random variable of type I, see (3.1.3). For $x \leq 0$ we suppose $\tilde{F}(x) = 0$.

Denote

$$F_*(x) = \inf_{x \leq t} F(t), \quad F^*(x) = \sup_{0 \leq t \leq x} F(t).$$

Uniform convergence in (2.1) yields

$$(F^*(a_n x))^n \to \tilde{F}(x), \quad (F_*(a_n x))^n \to \tilde{F}(x) \text{ as } n \to \infty \tag{2.2}$$

for each positive x. Since T is a Choquet capacity, the function F is lower semi-continuous. Hence the functions F^* and F_* are left-continuous. For example,

$$\begin{aligned} F^*(x) \geq F^*(x-0) &= \lim_{y \uparrow x} F^*(y) \\ &\geq \liminf_{y \to x} F(y) \geq F(x). \end{aligned}$$

Since $\Psi(K)$ is finite, $F^*(\infty) = F_*(\infty) = 1$, i.e. F^* and F_* are distribution functions, and K belongs to $\mathcal{T}$.

From (2.2) and the necessary conditions in limit theorems for maxima (Galambos, 1978) we derive that the functions $1 - F_*(x)$ and $1 - F^*(x)$ are regularly varying with exponent α.

Evidently,

$$1 - F^*(x) \leq 1 - F(x) \leq 1 - F_*(x).$$

Let $s > 1$ be specified, and let $n(k)$ be the integer part of s^k. Then, for all sufficiently large t, there exists k such that $a_{n(k)} \leq t < a_{n(k+1)}$ and also

$$\begin{aligned} F_*(a_{n(k)}) &\leq F_*(t) \leq F_*(a_{n(k+1)}), \\ F^*(a_{n(k)}) &\leq F^*(t) \leq F^*(a_{n(k+1)}). \end{aligned}$$

Hence

$$\frac{\log F_*(a_{n(k+1)})}{\log F^*(a_{n(k)})} \leq \frac{\log F_*(t)}{\log F^*(t)} \leq \frac{\log F_*(a_{n(k)})}{\log F^*(a_{n(k+1)})}.$$

It follows from (2.2) that

$$\frac{\log F_*(t)}{\log F^*(t)} \to 1 \text{ as } t \to \infty,$$

whence

$$\lim_{x \to \infty} \frac{1 - F_*(x)}{1 - F^*(x)} = 1.$$

This fact and the regular variation of F^* and F_* imply the regular variation of the function $\tau_K(x)$.

The dual case $\alpha > 0$ is considered similarly. It is reduced to the limit theorem for minima. □

Consider an example of a random set such that the capacity functional of the corresponding unions converges only pointwise.

EXAMPLE 2.2 Let $Y \subset (-\infty, 0]$ be a U-stable RACS with $\alpha = -1$, and let M_k be equal to $[k, k+1/2]$, $k > 1$, with probability k^{-3} and be empty otherwise. Furthermore, let $A_1, A_2, \ldots$ be iid copies of the random set A defined as

$$A = Y \cup M_1 \cup M_2 \cup \ldots$$

Put $a_n = n$. Then the capacity functional of $n^{-1}X_n$ is equal to

$$\begin{aligned} T_n(K) &= 1 - (1 - T(nK))^n (1 - T'(nK))^n \\ &= 1 - (1 - T(K))(1 - T'(nK))^n, \end{aligned}$$

where T, T' are the capacity functionals of Y and $M_1 \cup M_2 \cup \ldots$ respectively.

Let $0 \notin K \subseteq [a, b]$ for a certain positive a. Then

$$1 - T'(K) \geq \left(1 - [na]^{-3}\right)^{[n(b-a]},$$

where $[na]$ designates the integer part of na. Hence

$$(1 - T'(nK))^n \to 1 \text{ as } n \to \infty,$$

so that $T_n(K) \to T(K)$ as $n \to \infty$. Thus, $n^{-1}X_n$ converges weakly to Y as $n \to \infty$. However, corresponding capacity functionals do not converge uniformly on the family $\{xK, x > 0\}$ even for $K = \{1\}$. Indeed, $T(\{x\}) = 0$ for all $x > 0$, and also

$$\begin{aligned} \sup_{x>0} |T_n(\{x\}) - T(\{x\})| &\geq T_n(\{x_n\}) \\ &= 1 - \left(1 - [nx_n]^{-3}\right)^n \to 1 - e^{-1} \text{ as } n \to \infty \end{aligned}$$

for $x_n = [n^{1/3}]n^{-1}$, $n \geq 1$. Note that the function $T(xK)$ is not regularly varying for the given K.

It seems interesting to construct an example such that $T(xK)$ is regularly varying, but the uniform convergence fails. If there is no such example, then, perhaps, the capacity functionals converge uniformly in conditions of Theorems 1.1 and 1.3. In this case it is interesting to find out exact necessary and sufficient conditions for the pointwise convergence.

4.3 Limit Theorems for Normalized Convex Hulls.

First, consider pointwise convergence of inclusion functionals of *convex hull* of random sets. The limiting random set is necessary convex-stable (see Sections 1.4 and 3.3 for notations and definitions).

Let A_n, $n \geq 1$, be iid random sets with the inclusion functional

$$\text{t}(F) = \mathbf{P}\{A_1 \subset F\}, F \in \mathcal{C},$$

and let

$$Z_n = \text{conv}(A_1 \cup \cdots \cup A_n).$$

Note that the sets A_n, $n \geq 1$, are allowed to be non-convex and unbounded.

We consider limit theorems for the inclusion functional of $a_n^{-1} Z_n$ for a certain normalizing sequence a_n, $n \geq 1$, of positive real numbers. Evidently, for any convex F,

$$\mathfrak{t}_n(F) = \mathbf{P}\left\{a_n^{-1} Z_n \subset F\right\} = (\mathfrak{t}(a_n F))^n. \tag{3.1}$$

If $\mathfrak{t}_n(F)$ tends to $\tilde{\mathfrak{t}}(F)$ as $n \to \infty$ for each convex F, then the limiting functional $\tilde{\mathfrak{t}}$ (if non-trivial) is a capacity functional of a certain strictly C-stable random set Z. The limiting functional satisfies (3.3.8). In other words, Z is of the second type with $H = \{0\}$. Thus

$$\tilde{\mathfrak{t}}(F) = \exp\{\tilde{\psi}(F)\}, \quad F \in \mathcal{C}, \ F \supset F_A, \tag{3.2}$$

and, for a certain $\alpha \neq 0$,

$$\tilde{\psi}(sF) = s^\alpha \tilde{\psi}(F), \ s > 0. \tag{3.3}$$

The notations below resemble notations introduced in Section 4.1. Let $a_n \to +\infty$ as $n \to \infty$. Then the origin is a fixed point of the limiting random set, so that $\alpha < 0$ by Corollary 3.3.6. Define for any convex F

$$a_n(F) = \sup\{x \colon \mathfrak{t}(xF) \leq 1 - 1/n\}. \tag{3.4}$$

If $\mathfrak{t}(xF) > 1 - 1/n$ for all $x \geq 0$, then put $a_n(F) = 0$. Introduce the sub-family of convex sets by

$$\mathcal{T}_c = \left\{F \in \mathcal{C} \colon \limsup_{x \to \infty} \mathfrak{t}(xF) = 1\right\}.$$

The following theorem resembles Theorem 1.1, although its proof is simpler, since the limit theorem for convex hulls can be reduced directly to the limit theorem for extremes of random variables.

Theorem 3.1 *Let $F \in \mathcal{T}_c$. Suppose that there is*

$$\lim_{n \to \infty} a_n(F)/a_n = q(F),$$

which is allowed to be infinite. Then

$$\lim_{n \to \infty} \mathfrak{t}_n(F) = \exp\left\{-q(F)^{-\alpha}\right\}, \tag{3.5}$$

if (and only if in case $0 < q(F) < \infty$) the function

$$\tau_F(x) = 1 - \mathfrak{t}(xF), \ x \geq 0,$$

is regularly varying with exponent $\alpha < 0$. If $F \notin \mathcal{T}_c$ then $\mathfrak{t}_n(F) \to 0$ as $n \to \infty$.

PROOF. Evidently, $\mathfrak{t}_n(F) \to 0$ as $n \to \infty$ if $0 \notin F$. Let $0 \in F$. Put

$$\xi(A_i) = \inf\{s \geq 0 \colon A_i \subseteq sF\}.$$

The distribution function of the random variable $\xi(A_i)$ is evaluated as

$$\begin{aligned} F_\xi(x) &= \mathbf{P}\{\xi(A_i) < x\} \\ &= \mathbf{P}\{A \subseteq xF\} = \mathfrak{t}(xF). \end{aligned}$$

Hence
$$a_n(F) = \sup\{x\colon F_\xi(x) \le 1 - 1/n\},$$
and F belongs to $\mathcal{T}_c$ if and only if $F_\xi(x) \to 1$ as $x \to \infty$. Moreover,

$$\zeta_n = \xi(\mathrm{conv}(A_1 \cup \dots \cup A_n)) = \max\{\xi(A_i), 1 \le i \le n\}.$$

As in Galambos (1978), we get

$$\mathbf{P}\{\zeta_n < a_n(F)x\} \to \exp\{-x^\alpha\} \quad \text{as } n \to \infty.$$

Then (3.5) is valid, since

$$\begin{aligned} \mathfrak{t}_n(a_n F) &= \mathbf{P}\{\zeta_n < a_n\} \\ &= \mathbf{P}\left\{\zeta_n < a_n(F)\frac{a_n}{a_n(F)}\right\}. \end{aligned}$$

If $0 < q(F) < \infty$, then the necessity follows from the corresponding theorem for extremes of random variables, see Galambos (1978).

It is easy to show that for any $F \in \mathcal{T}_c$ and $s > 0$ the set sF belongs to $\mathcal{T}_c$ and $\tilde{\psi}(sF) = s^\alpha \tilde{\psi}(F)$, so that the limiting distribution corresponds to a certain C-stable set. □

It was proven in Section 1.4 that the pointwise convergence of inclusion functionals implies the weak convergence of convex compact random sets. Thus, the following theorem is valid.

Theorem 3.2 *Let the conditions of Theorem 3.1 be valid for each F from $\mathcal{T}_c \cap \mathcal{C}_0$, and let the random set A_1 be compact almost surely. Then $a_n^{-1} Z_n$ converges weakly to the C-stable set Z with the inclusion functional $\tilde{\mathfrak{t}}$ given by*

$$\tilde{\mathfrak{t}}(F) = \begin{cases} \exp\{-q(F)^\alpha\} & , \quad F \in \mathcal{T}_c \\ 0 & , \quad \text{otherwise} \end{cases}. \tag{3.6}$$

In fact, the class $\mathcal{C}_0$ in Theorem 3.2 can be replaced with the class $\mathcal{P}$ of all bounded convex polyhedrons in $\mathbb{R}^d$, see Proposition 1.4.4.

Now consider a limit theorem for the Aumann expectation of convex compact sets (see Section 2.1 for the definition of expectation).

Theorem 3.3 *Let $a_n^{-1} Z_n$ converge weakly to the random set Z. Then, for any $R > 0$, the random set $a_n^{-1} Z_n \cap B_R(0)$ converges weakly to $Z \cap B_R(0)$, and also*

$$\mathbf{E}\left[a_n^{-1} Z_n \cap B_R(0)\right] \to \mathbf{E}\left[Z \cap B_R(0)\right] \quad \text{as } n \to \infty$$

in the Hausdorff metric. Moreover,

$$a_n^{-d} \mu\left(\mathbf{E}\left[Z_n \cap a_n B_R(0)\right]\right) \to \mu\left(\mathbf{E}\left[Z \cap B_R(0)\right]\right) \quad \text{as } n \to \infty, \tag{3.7}$$

where μ is the Lebesgue measure in $\mathbb{R}^d$.

PROOF is straightforward and simply follows from Theorem 1.4.6. □

We provide only the outline of the case of convergence to the limiting functional with $\alpha > 0$. Let $a_n \to 0$ as $n \to \infty$. Define for any F belonging to $\mathcal{C}$

$$\bar{a}_n(F) = \inf\{x > 0 \colon \mathfrak{t}(xF) \leq 1 - 1/n\}.$$

If $\mathfrak{t}(xF) > 1 - 1/n$ for all $x \geq 0$, then put $\bar{a}_n(F) = \infty$. Let

$$\bar{\mathcal{T}}_c = \left\{F \in \mathcal{C} \colon \liminf_{x \to 0} \mathfrak{t}(xF) = 1\right\}.$$

Theorem 3.4 *Let $F \in \bar{\mathcal{T}}_c$. Suppose that there is*

$$\lim_{n\to\infty} \bar{a}_n(F)/a_n = \bar{q}(F),$$

which is allowed to be infinite. Then

$$\lim_{n\to\infty} \mathfrak{t}_n(F) = \exp\left\{-\bar{q}(F)^{-\alpha}\right\}, \tag{3.8}$$

if (and only if in case $0 < \bar{q}(F) < \infty$) the function $\tau_F(x) = 1 - \mathfrak{t}(xF)$, $x \geq 0$, is regularly varying at zero with exponent $\alpha > 0$. If $F \notin \bar{\mathcal{T}}_c$, then $\mathfrak{t}_n(F) \to 0$ as $n \to \infty$.

Note that the limiting functional in Theorem 3.4 corresponds to necessary unbounded random set, so that (3.8) cannot be reformulated directly in terms of the weak convergence.

Naturally, limit theorems for convex hulls follow from the corresponding results for unions. However, to prove the convergence of unions we have to check regular variation conditions for all compacts. This is sometimes more tiresome than to check conditions of Theorem 3.1. Besides, for the convergence of convex hulls the necessary and sufficient conditions have been obtained.

4.4 Limit Theorems for Unions and Convex Hulls of Special Random Sets

First, consider convergence of *random samples* in $\mathbb{R}^d$ and their convex hulls. In this case $A_1 = \{\xi_1\}$ is a single-point random set and $X_n = \{\xi_1, \ldots, \xi_n\}$ for iid random vectors $\xi_1, \ldots, \xi_n$. We shall prove that the random set $a_n^{-1} X_n$ admits a non-trivial weak limit if the random vector ξ has a regularly varying density.

Hereafter in this section the numerical function $f \colon \mathbb{R}^d \to \mathbb{R}$ is called *regularly varying* if f belongs to the class Π_2 (see Section 1.6). It means that

$$\sup_{u \in \mathbb{S}^{d-1}} \left|\frac{f(xu)}{f(xe)} - \phi_e(u)\right| \to 0 \text{ as } x \to \infty \tag{4.1}$$

for any vector e from $\mathbb{R}^d \setminus \{0\}$. To make this section more self-contained recall that any regularly varying function f admits the representation

$$f(u) = \phi(u)L(u), \tag{4.2}$$

where L is a slowly varying function and $\phi(u)$ is a continuous homogeneous function.

Theorem 4.1 *Let $A_1 = M + \xi$, where ξ is a random vector with a regularly varying positive density f, $\operatorname{ind} f = \alpha - d$, $\alpha < 0$, and M is a non-empty RACS independent of ξ. Suppose that $M \subset K_0$ a.s. for a certain compact K_0. Furthermore, let ϕ and L be the corresponding factors in (4.2). Put*

$$a_n = \sup\{x\colon\ x^\alpha L(xe) \geq 1/n\} \tag{4.3}$$

for a certain e belonging to $\mathbb{R}^d \setminus \{0\}$. Then $a_n^{-1}X_n = a_n^{-1}(A_1 \cup \cdots \cup A_n)$ converges weakly to the U-stable compact RACS X with the capacity functional

$$\tilde{T}(K) = 1 - \exp\left\{-\int_K \phi(u)du\right\}. \tag{4.4}$$

PROOF. Evidently, $a_n \to \infty$ as $n \to \infty$. Check the conditions of Theorem 1.1 on the class $\mathcal{K}_{ub}$. Let $K \in \mathcal{K}_{ub}$, and let $0 \in K$. Then there exists a ball K_1 such that $0 \in K_1 \subset K$. Since f is positive, we get

$$\begin{aligned} \tau_K(x) = T(xK) &\geq \mathbf{P}\{M + \xi \cap xK_1 \neq \emptyset\} \\ &= \mathbf{P}\{\xi \in xK_1 \oplus \check{M}\} \\ &\geq \mathbf{P}\{\xi \in K_1 \oplus \check{M}\} > 0, \end{aligned}$$

where $\check{M} = \{-x\colon x \in M\}$. Thus, $K \notin \mathcal{T}$, and

$$\mathbf{P}\left\{a_n^{-1}X_n \cap K \neq \emptyset\right\} \to 1 \text{ as } n \to \infty.$$

Let $K \in \mathcal{K}_{ub}$, and, moreover, $0 \notin K^\varepsilon$ for a certain $\varepsilon > 0$. Then

$$\begin{aligned} \tau_K(x) &= \mathbf{P}\{\xi \in xK \oplus \check{M}\} \\ &\leq \int_{xK \oplus \check{K}_0} f(u)du \\ &= x^d \int_{K \oplus \check{K}_0/x} f(xu)du \\ &= x^\alpha \int_{K^\varepsilon} \phi(u)L(xu)du. \end{aligned}$$

for sufficiently large x.

It follows from (1.6.9) (see also Lemma 6.3.2) that

$$\int_K \phi(u)L(xu)du \sim L(xe)\int_K \phi(u)du \text{ as } x \to \infty.$$

Thus

$$\tau_K(x) \leq x^\alpha L(xe)(1 + \lambda_x)\Lambda(K^\varepsilon), \tag{4.5}$$

where $\lambda_x \to 0$ as $x \to \infty$ and

$$\Lambda(K) = \int_K \phi(u)du, \; K \in \mathcal{K}. \tag{4.6}$$

It follows from the theory of regularly varying functions that $x^\alpha L(xe) \to 0$ as $x \to \infty$ for $\alpha < 0$, see Section 1.6. Hence $\tau_K(x) \to 0$ as $x \to \infty$, whence it follows that K belongs to the class $\mathcal{T}$ defined in (1.4).

Estimate $\tau_K(x)$ from below in the similar way

$$\begin{aligned}\tau_K(x) &\geq \inf_{x \in K_0} x^\alpha \int_{K-y/x} \phi(u)L(xu)du \\ &\geq \inf_{y \in B_\varepsilon(0)} x^\alpha \int_{K-y} \phi(u)L(xu)du.\end{aligned}$$

Since $K \in \mathcal{K}_{ub}$, we get

$$\bigcap_{y \in B_\varepsilon(0)} (K-y) = K^{-\varepsilon} = \{y\colon\ B_\varepsilon(y) \subseteq K\}\,.$$

Hence, in the above introduced notations,

$$\tau_K(x) \geq x^\alpha L(xe)(1-\lambda_x)\Lambda(K^{-\varepsilon}).$$

Thus, for all K belonging to the class $\mathcal{K}_{ub}$ of finite unions of balls,

$$x^\alpha \frac{L(txe)(1-\lambda_{xt})\Lambda(K^{-\varepsilon})}{L(te)(1+\lambda_t)\Lambda(K^\varepsilon)} \leq \frac{\tau_K(xt)}{\tau_K(t)} \leq x^\alpha \frac{L(txe)(1+\lambda_{xt})\Lambda(K^\varepsilon)}{L(te)(1-\lambda_t)\Lambda(K^{-\varepsilon})},$$

provided K misses the origin. Hence

$$x^\alpha \frac{\Lambda(K^{-\varepsilon})}{\Lambda(K^\varepsilon)} \leq \lim_{t\to\infty} \frac{\tau_K(xt)}{\tau_K(t)} \leq x^\alpha \frac{\Lambda(K^\varepsilon)}{\Lambda(K^{-\varepsilon})}.$$

Since $K \in \mathcal{K}_{ub}$, we get $K^{-\varepsilon} \uparrow \mathrm{Int}K$ as $\varepsilon \downarrow 0$. The continuity of ϕ yields $\Lambda(K^\varepsilon) \downarrow \Lambda(K)$ and $\Lambda(K^{-\varepsilon}) \uparrow \Lambda(K)$ as $\varepsilon \downarrow 0$. Hence

$$\lim_{t\to\infty} \frac{\tau_K(xt)}{\tau_K(t)} = x^\alpha,$$

i.e. the function $\tau_K(x)$ is regularly varying with exponent α.

It follows from (1.2), (4.5) that for all $\beta > 0$ and sufficiently large n

$$\begin{aligned}a_n(K) &\leq \sup\{x\colon\ x^\alpha L(xe)(1+\beta)\Lambda(K^\varepsilon) \geq 1/n\} \\ &= \sup\{x\colon\ s(x) \leq n\Lambda(K^\varepsilon)(1+\beta)\}\,,\end{aligned}$$

where $s(x) = x^{-\alpha}L(xe)$ is a regularly varying function such that $\mathrm{ind}s = -\alpha$. According to Seneta (1976), $s(x)$ admits the asymptotic inverse function $\bar{s}(x)$, which is regular varying with

$$\mathrm{ind}\bar{s} = \gamma = -\frac{1}{\alpha}.$$

Then

$$\lim_{n\to\infty} \frac{a_n(K)}{\bar{s}(n\Lambda(K^\varepsilon(1+\beta))} \leq 1,$$

and, by (4.3),

$$a_n = \sup\{x\colon\ s(x) \leq n\} \sim \bar{s}(n) \ \text{ as } \ n \to \infty.$$

Theorem 1.1 yields

$$\Psi(K) \leq \lim_{n\to\infty} \left(\frac{\bar{s}(n\Lambda(K^\varepsilon)(1+\beta))}{\bar{s}(n)}\right)^{-\alpha} = \Lambda(K^\varepsilon)(1+\beta).$$

Similarly,

$$\Psi(K) \geq \Lambda(K^{-\varepsilon})(1-\beta).$$

Letting ε go to zero yields $\Psi(K) = \Lambda(K)$, whence (4.5) is valid for each K from $\mathcal{K}_{ub}$. It follows from the general results on distributions of random sets (see Matheron, 1975) that (4.5) is valid for each compact K (if $0 \in K$ we assume $\int_K \phi(u)du = \infty$). □

NOTE. If $f(xu) \sim \phi(xu)$ as $x \to \infty$, for a certain homogeneous function ϕ and any u from $\mathbb{R}^d \setminus \{0\}$, then the statement of Theorem 4.1 is valid for $a_n = n^\gamma$, $\gamma = -1/\alpha$.

The limiting random set X in Theorem 4.1 is the Poisson point process in $\mathbb{R}^d$ with the intensity measure Λ given by (4.6). Its distribution does not depend on the shape of M, provided M is contained a.s. inside a certain compact.

The following theorem deals with the convergence of convex hulls of random samples.

Theorem 4.2 *Let the conditions of Theorem 4.1 be valid, and let*

$$Z_n = \text{conv}(A_1 \cup \cdots \cup A_n).$$

Then $a_n^{-1}Z_n$ converges weakly to $Z = \text{conv}(X)$, where X is the weak limit of $a_n^{-1}(A_1 \cup \cdots \cup A_n)$. The inclusion functional of the limiting random set Z is defined as

$$\tilde{\imath}(F) = \exp\left\{-\int_{F^c} \phi(u)du\right\}. \tag{4.7}$$

PROOF follows from Theorem 4.1, since for any convex F

$$\mathbf{P}\{Z \subseteq F\} = \mathbf{P}\{X \cap F^c = \emptyset\} = 1 - T(F^c).$$

The limiting random set Z is strictly C-stable with $\gamma = -1/\alpha > 0$ and $H = \{0\}$ (see Theorem 3.3.5). □

It was proven in Vitale (1987) that expectations of convex hulls for n iid random vectors, $n \geq 1$, determine uniquely the distribution of the random vector in question. Let us proceed to evaluate the expectation of the limiting convex random set Z in Theorem 4.2.

Theorem 4.3 *If $\alpha < -1$, then the expectation of the C-stable set Z with the inclusion functional (4.7) is the convex compact set $\mathbf{E}Z$ having the support function*

$$s_{\mathbf{E}Z}(v) = \Gamma(1-\gamma)\left[\gamma \int_{S_v^+} \phi(u)(u \cdot v)^{1/\gamma}du\right]^\gamma, v \in \mathbb{S}^{d-1}, \tag{4.8}$$

where $S_v^+ = \{u \in \mathbb{S}^{d-1} : (u \cdot v) \geq 0\}$, $\gamma = -1/\alpha$, Γ is the gamma-function.

PROOF. It is obvious that

$$\mathbf{P}\{s_Z(v) < x\} = \mathbf{P}\{A \subseteq xH_v^c\} = \exp\{-x^\alpha a(v)\}$$

where $H_v = \{u : u \cdot v \geq 1\}$ is the half-space touching the unit sphere at the point v and $a(v) = \int_{H_v} \phi(w)dw$. Ordinary evaluations yield

$$\begin{aligned} s_{\mathbf{E}Z}(v) &= \mathbf{E}s_Z(v) = a(v)^{-1/\alpha}\Gamma(1/\alpha + 1) \\ &= a(v)^{\gamma}\Gamma(1-\gamma). \end{aligned} \tag{4.9}$$

Let $w = uy$, for $u \in S_v^+$, and let $y \geq 1/(u \cdot v)$. Thus, $dw = y^{d-1}dudy$ and

$$\begin{aligned} a(v) &= \int_{S_v^+} du \int_{(u\cdot v)^{-1}}^{\infty} \phi(yu)y^{d-1}dy \\ &= \int_{S_v^+} du \int_{(u\cdot v)^{-1}}^{\infty} y^{\alpha-d}\phi(u)y^{d-1}dy \\ &= -\alpha^{-1}\int_{S_v^+} \phi(u)(u\cdot v)^{-\alpha}du. \end{aligned}$$

Now (4.8) follows from (4.9). Note that integrals over any part of $\mathbb{S}^{d-1}$ are understood with respect to the $(d-1)$-dimensional Lebesgue measure on $\mathbb{S}^{d-1}$. □

The expectation $\mathbf{E}Z$ may be used in statistics for testing for lack of circular symmetry for random samples.

EXAMPLE 4.4 Let the function ϕ be spherically symmetric, i.e. $\phi(u) = C$ for all $u \in \mathbb{S}^{d-1}$. Then the inclusion functional of Z is equal to

$$\tilde{\imath}(F) = \exp\left\{-C\int_{F^c} \|u\|^{\alpha-d}du\right\},$$

and $\mathbf{E}Z$ is the ball $B_r(0)$ for r given by

$$r = \Gamma(1-\gamma)\left[\gamma C\int_{S_v^+}(u\cdot v)^{1/\gamma}du\right]^{\gamma}. \tag{4.10}$$

Similarly to the expectation of the random set Z, the expectation of its norm $\|Z\| = \sup\{\|z\| : z \in Z\}$ is evaluated as

$$\mathbf{E}\|Z\| = \Gamma(1-\gamma)\left[\gamma s_{d-1}\int_{\mathbb{S}^{d-1}} \phi(u)\right]^{\gamma},$$

where s_{d-1} is the surface area of the unit sphere in $\mathbb{R}^d$.

If ϕ is spherically symmetric, then

$$\frac{\|\mathbf{E}Z\|}{\mathbf{E}\|Z\|} = \left[\int_{S_v^+}(u\cdot v)^{1/\gamma}du/s_{d-1}\right]^{-\gamma}.$$

In particular, for $d = 2$ it is

$$\begin{aligned} \frac{\|\mathbf{E}Z\|}{\mathbf{E}\|Z\|} &= \left[\frac{1}{\pi}\int_0^{\pi/2}(\cos\beta)^{1/\gamma}d\beta\right]^{-\gamma} \\ &= \left[\frac{\Gamma((1-\alpha)/2)}{2\pi^{1/2}\Gamma(1-\alpha/2)}\right]^{1/\alpha}. \end{aligned}$$

Applications of the results above will be discussed in Section 8.2. Note only that from Theorems 4.2 and 4.3 estimates for tail probabilities for the volume of convex hulls of random samples can be derived.

As it was stated above, the limiting distribution of $a_n^{-1}X$ in the scheme of Theorem 4.1 does not depend on the shape of M, provided M is contained almost surely in a certain compact. Otherwise the limiting distribution becomes more complicated.

The following theorem deals with iid copies of the random closed set defined as $A_1 = M(\xi)$, where $M: \mathbb{R}^m \to \mathcal{F}$ is a multivalued function and ξ is a random vector in $\mathbb{R}^m$ having a regularly varying density. Suppose that

$$\sup\{\|M(u)\|:\ u \in \mathbb{S}^{m-1}\} < \infty.$$

Theorem 4.5 *Let ξ be a random vector in $\mathbb{R}^m$ having positive regularly varying density f, $\operatorname{ind} f = \alpha - m$, $\alpha < 0$, and let $M: \mathbb{R}^m \to \mathcal{K}$ be a homogeneous set-valued function whose values are compact convex subsets of $\mathbb{R}^d$, i.e.*

$$M(xu) = x^\eta M(u) \tag{4.11}$$

for a certain $\eta > 0$, whatever $x > 0$ and $u \in \mathbb{R}^m$ may be. Furthermore, let $A_1 = M(\xi)$ be a random closed set. For a certain vector $e \in \mathbb{R}^m \setminus \{0\}$ denote

$$\mathcal{L}_K = \{u \in \mathbb{R}^m:\ M(u) \cap K \neq \emptyset\}, \tag{4.12}$$

$$a_n = \sup\{x^\eta:\ x^\alpha L(xe) \geq 1/n\}, \tag{4.13}$$

where $f = \phi L$ for a slowly varying function L and homogeneous ϕ. Then the random closed set

$$a_n^{-1} X_n = a_n^{-1}(A_1 \cup \cdots \cup A_n)$$

converges weakly to the U-stable set X with the capacity functional $\tilde{T}$ defined as

$$\tilde{T}(K) = \begin{cases} 1 - \exp\left\{-\int_{\mathcal{L}_K} \phi(u)du\right\}, & 0 \notin K \\ 1, & \text{otherwise} \end{cases}. \tag{4.14}$$

PROOF. It is obvious that

$$\mathcal{L}_{xK} = x^{1/\eta}\mathcal{L}_K$$

for all $x > 0$ and $K \in \mathcal{K}$. Note that $0 \in \mathcal{L}_K$ as long as $0 \in K$. In the above introduced notations we get

$$\begin{aligned} \tau_K(x) &= \mathbf{P}\{\xi \in \mathcal{L}_{xK}\} = \mathbf{P}\{\xi \in x^{1/\eta}\mathcal{L}_K\} \\ &= x^{\alpha/\eta}\int_{\mathcal{L}_K} \phi(u) L(x^{1/\eta}u)du. \end{aligned}$$

If $0 \notin K$, then $\tau_K(x) \to 0$ as $x \to \infty$ and

$$x^{\alpha/\eta}L(x^{1/\eta}e)(1-\lambda_x)\Lambda(\mathcal{L}_K) \leq \tau_K(x) \leq x^{\alpha/\eta}L(x^{1/\eta}e)(1+\lambda_x)\Lambda(\mathcal{L}_K).$$

Hence $\tau_K(x)$ is regularly varying with exponent α/η. The proof is completed similarly to the proof of Theorem 4.1. □

NOTE. The dimension m is allowed to be different from d. If the vector ξ is distributed in a certain cone $\mathbb{C} \subset \mathbb{R}^m$, and $M: \mathbb{C} \to \mathcal{K}$, then (4.14) remains true for

$$\mathcal{L}_K = \{u \in \mathbb{C}:\ M(u) \cap K \neq \emptyset\}.$$

EXAMPLE **4.6** Let $m = d = 1$, and let ξ be a random variable having Cauchy distribution. Furthermore, let $A_1 = \xi M$ be a random subset of $\mathbb{R}$, where M is a non-random compact missing the origin. Then $a_n = n$, and the random set $n^{-1}(\xi_1 M \cup \dots \cup \xi_n M)$ converges weakly to the U-stable random set X with $\alpha = -1$. The capacity functional of X is evaluated as

$$\tilde{T}(K) = 1 - \exp\left\{-\int_{K/M} u^{-2} du\right\}.$$

where $K/M = \{x/y : x \in K, y \in M\}$. In particular, for $M = \{1\}$ the random sample $n^{-1}\{\xi_1, \dots, \xi_n\}$ converges weakly to the random set X with the capacity functional given by

$$\tilde{T}(K) = 1 - \exp\left\{-\int_K u^{-2} du\right\}.$$

Clearly, X is the Poisson point process with the intensity function u^{-2}.

Then consider convergence of convex hulls of special random closed sets. Similar to Theorem 4.2, we obtain the following result.

Theorem 4.7 *Let the conditions of Theorem 4.5 be valid. Then*

$$a_n^{-1} Z_n = a_n^{-1} \mathrm{conv}(A_1 \cup \dots \cup A_n)$$

converges weakly to the C-stable random closed set Z with the inclusion functional

$$\tilde{\imath}(F) = exp\left\{-\int_{\{u:\, M(u) \not\subset F\}} \phi(u) du\right\}, F \in \mathcal{C}. \tag{4.15}$$

If K is a convex compact set and $0 \in \mathrm{Int} K$, *then*

$$\tilde{\imath}(K) = \exp\left\{\frac{1}{\alpha}\int_{\mathbb{S}^{d-1}} \left[\inf_{u \in \mathbb{S}^{m-1}, s_{M(e)}(u) > 0} \frac{s_K(u)}{s_{M(e)}(u)}\right]^{\alpha/\eta} \phi(e) de\right\}, \tag{4.16}$$

where $s_K(\cdot)$, $s_{M(e)}(\cdot)$ *are the support functions of K and M(e).*

If $\alpha/\eta < -1$, *then the expectation* $\mathbf{E}Z$ *of the limiting convex RACS Z exists and has the support function*

$$s_{\mathbf{E}Z}(v) = \Gamma(1 - \gamma\eta)\left[\gamma\eta \int_{\{u:\, M(u) \cap H_v \neq \emptyset\}} \phi(u) du\right]^{\gamma\eta}, v \in \mathbb{S}^{d-1},$$

where $H_v = \{u \in \mathbb{R}^d : u \cdot v \geq 1\}$, $\gamma = -1/\alpha$.

PROOF. Evidently, (4.15) follows from (4.14). Then

$$\begin{aligned} \{u:\, M(u) \not\subset K\} &= \{xe:\, x^\eta M(e) \not\subset K\} \\ &= \{xe:\, x^\eta \geq h_K(e)\}, \end{aligned}$$

where

$$h_K(e) = \inf\left\{\frac{s_K(u)}{s_{M(e)}(u)} : u \in \mathbb{S}^{m-1}, s_{M(e)}(u) > 0\right\}.$$

Thus

$$\int_{\{u:\, M(u)\not\subset F\}} \phi(u)du = \int_{\mathbb{S}^{m-1}} \phi(e)de \int_{[h_K(e)]^{1/\eta}}^{\infty} x^{\alpha-1}dx$$
$$= -\frac{1}{\alpha}\int_{\mathbb{S}^{m-1}} h_K(e)^{\alpha/\eta}\phi(e)de.$$

Then (4.16) follows from (4.15). The evaluation of the support function of $\mathbf{E}Z$ is straightforward. □

Note that Theorem 4.5 and 4.7 make it possible to obtain limit theorems for unions and convex hulls of random balls ($m = d+1$, $M(u_1,\ldots,u_{d+1})$ is the ball in $\mathbb{R}^d$ of radius u_{d+1} centered at $(u_1,\ldots,u_d)$) or random triangles ($m = 3d$, and $M(u_1,\ldots,u_{3d})$ is the triangle with the vertices $(u_1,\ldots,u_d)$, $(u_{d+1},\ldots,u_{2d})$, $(u_{2d+1},\ldots,u_{3d})$) etc. In these cases $M(su) = sM(u)$ for all $u \in \mathbb{R}^m$ and $s > 0$, whence $\eta = 1$.

Consider a consequence from Theorems 4.5 and 4.7.

Theorem 4.8 *Let $(\xi_1,\ldots,\xi_d,\zeta)$ be a random vector in $\mathbb{R}^d \times [0,\infty)$ with the regularly varying density $f(u;y)$, $\operatorname{ind} f = \alpha - d - 1$, $\alpha < 0$, and let $A_1 = \xi + \zeta M$, where $\xi = (\xi_1,\ldots,\xi_d)$, $M \in \mathcal{K}$, $0 \in M$. Furthermore, let*

$$f(u;y) = \phi(u;y)L(u;y),\ u \in \mathbb{R}^d,\ y > 0,$$

where ϕ is homogeneous and L is slowly varying. Put

$$a_n = \sup\{x\colon\ x^\alpha L(xe;xt) \ge 1/n\}$$

for a certain point $(e;t)$ from $(\mathbb{R}^d\setminus\{0\})\times(0,\infty)$. Then $a_n^{-1}X_n$ converges weakly to the U-stable RACS X with the capacity functional

$$\tilde{T}(K) = \exp\left\{-\int_0^\infty dy \int_{K\oplus y\check{M}} \phi(u;y)du\right\}. \tag{4.17}$$

Moreover, $a_n^{-1}Z_n = a_n^{-1}\operatorname{conv}(X_n)$ converges weakly to the C-stable RACS Z with the inclusion functional

$$\tilde{\imath}(F) = \exp\left\{-\int_0^\infty dy \int_{F^c\oplus y\check{M}} \phi(u;y)du\right\}. \tag{4.18}$$

If $\gamma = -1/\alpha < 1$, then the expectation $\mathbf{E}Z$ has the support function

$$s_{\mathbf{E}Z}(v) = \Gamma(1-\gamma)\left[\gamma\int_0^\infty dy \int_{S_v^+} \phi(w;y)\left((w\cdot v) + ys_M(v)\right)^{1/\gamma} dw\right]^\gamma. \tag{4.19}$$

PROOF. Formulae (4.17) and (4.18) simply follow from (4.14) and (4.15) for $\eta = 1$. Theorem 4.7 yields

$$s_{\mathbf{E}Z}(v) = \Gamma(1-\gamma)\left[\gamma\int_{F_v} \phi(u;y)dudy\right]^\gamma,\ v \in \mathbb{S}^{d-1}, \tag{4.20}$$

where

$$F_v = \left\{(u_1,\ldots,u_d,y)\colon\ u \in \mathbb{R}^d, y \ge 0, M(u,y)\cap H_v \ne \emptyset\right\},$$

$$M(u,y) = u + yM,$$

and $u = (u_1, \ldots, u_d)$. Hence

$$\begin{aligned} F_v &= \{(u_1, \ldots, u_d, y)\colon s_{u+yM}(v) \geq 1\} \\ &= \{(u_1, \ldots, u_d, y)\colon u \cdot v \geq 1 - y s_M(v)\}. \end{aligned}$$

Similar to the proof of Theorem 4.3 we get

$$\begin{aligned} a(v) &= \int_{F_v} \phi(u;y) du dy \\ &= \int_0^\infty dy \int_{S_v^+} dw \int_{(1-ys_M(v))/(w\cdot v)}^\infty \phi(rw;y) r^{d-1} dr. \end{aligned}$$

Let $y = y_1 r$. Then

$$r \geq \frac{1 - y_1 r s_M(v)}{w \cdot v},$$

whence

$$r \geq ((w \cdot v) + y_1 s_M(v))^{-1}.$$

Hence

$$\begin{aligned} a(v) &= \int_0^\infty r dy_1 \int_{S_v^+} dw \int_{((w\cdot v)+y_1 s_M(v))^{-1}} \phi(w; y_1) r^{\alpha-1} dr \\ &= \gamma \int_0^\infty dy \int_{S_v^+} \phi(w;y) ((w \cdot v) + y s_M(v))^{1/\gamma} dw. \end{aligned}$$

Now (4.19) follows from (4.20). □

Corollary 4.9 *Let the conditions of Theorem 4.8 be valid, and let M be a random compact set. Then the statements of Theorem 4.8 remain valid with*

$$\begin{aligned} \tilde{T}(K) &= 1 - \exp\left\{ -\int_0^\infty dy \mathbf{E}\left[\int_{K \oplus y\check{M}} \phi(u;y) du \right] \right\}, \\ \tilde{\imath}(F) &= \exp\left\{ -\int_0^\infty dy \mathbf{E}\left[\int_{F^c \oplus y\check{M}} \phi(u;y) du \right] \right\}, \\ s_{\mathbf{E}Z}(v) &= \Gamma(1-\gamma) \left\{ \gamma \int_0^\infty dy \int_{S_v^+} \phi(w;y) \mathbf{E}\left[((w\cdot v) + y s_M(v))^{1/\gamma} \right] dw \right\}^\gamma, \end{aligned}$$

instead of (4.17), (4.18), (4.19) respectively.

Let the conditions of Theorem 4.8 be valid, and let ξ have a spherically symmetric distribution. Then $\phi(w,y) = \bar{\phi}(y)$ for any w from $\mathbb{S}^{d-1}$. Therefore, (4.19) yields

$$s_{\mathbf{E}Z}(v) = \Gamma(1-\gamma) \left[\gamma \int_0^\infty \bar{\phi}(y) dy \int_{S_v^+} ((w\cdot v) + y s_M(v))^{1/\gamma} dw \right]^\gamma.$$

If $\gamma \uparrow 1$, then the support function $s_{EZ}(v)/\Gamma(1-\gamma)$ can be approximated with

$$c_1 \int_0^\infty \bar{\phi}(y) dy + c_2 \int_0^\infty y \bar{\phi}(y) dy s_M(v),$$

where c_2 is the surface area of S_v^+ and

$$c_1 = \int_{S_v^+} (w \cdot v) dw.$$

Roughly speaking, $\mathbf{E}Z/\Gamma(1-\gamma)$ can be approximated with the set $B_r(0) \oplus cM$, where

$$\begin{aligned} r &= c_1 \int_0^\infty \bar{\phi}(y) dy, \\ c &= c_2 \int_0^\infty y\bar{\phi}(y) dy. \end{aligned}$$

Thus, the set $\mathbf{E}Z/\Gamma(1-\gamma)$ inherits the shape of M as $\gamma \uparrow 1$.

EXAMPLE 4.10 Let $d = 2$, $\alpha = -2$, and let ξ have circular symmetric distribution. Furthermore, put $\phi(w; y) = (1 + y^4)^{-1}$ for each w from $\mathbb{S}^{d-1}$. It follows from (4.19) that the support function of $\mathbf{E}Z$ is equal to

$$s_{\mathbf{E}Z}(v) = \frac{\pi}{2^{5/4}} \left(\frac{\pi}{2} + 2^{3/2} s_M(v) + s_M(v)^2 \right)^{1/2}.$$

Below we provide only the outline of analogues of the previous results for convergence to U-stable and C-stable random sets with $\gamma = -1/\alpha < 0$.

Theorem 4.11 *Let $A_1 = \{\xi\}$ be a single-point random set, and let ξ be distributed with the density g in a certain convex cone $\mathbb{C} \subseteq \mathbb{R}^d$ which does not contain any half-space. Suppose that the function $f(u) = g(u\|u\|^{-2})$, $u \neq 0$, is regularly varying, $\operatorname{ind} f = d - \alpha$, $\alpha > 0$, with ϕ and L as factors in (4.2). For a certain $e \in \mathbb{C} \setminus \{0\}$ put*

$$a_n = \inf \{x > 0 \colon x^\alpha L(e/x) \geq 1/n\}. \tag{4.21}$$

Then $a_n^{-1} X_n = a_n^{-1}\{\xi_1, \ldots, \xi_n\}$ converges weakly to the U-stable set X with the capacity functional

$$\tilde{T}(K) = 1 - \exp\left\{-\int_{\mathbb{C}\cap K} \phi(u\|u\|^{-2}) du\right\}, \quad K \in \mathcal{K}. \tag{4.22}$$

The normalized convex hull $a_n^{-1} Z_n = a_n^{-1} \operatorname{conv}\{\xi_1, \ldots, \xi_n\}$ converges weakly to the strictly C-stable random set Z with parameters $\gamma = -1/\alpha < 0$, $H = \{0\}$ and the inclusion functional

$$\tilde{\imath}(F) = \exp\left\{-\int_{\mathbb{C}\setminus F} \phi(u\|u\|^{-2}) du\right\}, \quad F \in \mathcal{C}. \tag{4.23}$$

If the function $\rho(w) = \inf\{y \colon yw \in F\}$ is finite for each w from $\mathbb{S}^{d-1} \cap \mathbb{C}$, then

$$\tilde{\imath}(F) = \exp\left\{-\frac{1}{\alpha} \int_{\mathbb{C}\cap\mathbb{S}^{d-1}} \phi(w)\rho(w)^\alpha dw\right\}. \tag{4.24}$$

It should be noted that the limiting random set Z is unbounded almost surely, so that the expectation $\mathbf{E}Z$ does not exist. However, $\mathbf{E}s_Z(v)$ may be finite for some v.

Corollary 4.12 *Let $H_v(x) = \{u: u \cdot v \geq x\}$. Then, for any v from $\mathbb{C} \cap \mathbb{S}^{d-1}$,*

$$\begin{aligned}\mathfrak{t}(H_v(x)) &= \exp\{-x^\alpha a(v)\}, \\ \mathbf{E}s_Z(-v) &= -\Gamma(1+1/\alpha)a(v)^{-1/\alpha},\end{aligned}$$

where

$$a(v) = \frac{1}{\alpha} \int_{\mathbb{C} \cap \mathbb{S}^{d-1}} \phi(w)(w \cdot v)^{-\alpha} dw.$$

4.5 Further Remarks and Open Problems.

Consider a general normalization scheme for unions of random sets. Let X_n be the union of iid random sets $A_1, \dots, A_n$. For a sequence $a_n = (a_{n1}, \dots, a_{nd})$ of points belonging to $\mathbb{R}_+^d = [0, \infty)^d$, $n \geq 1$ put

$$a_n^{-1} \circ X_n = \left\{(a_{n1}^{-1} x_1, \dots, a_{nd}^{-1} x_d): (x_1, \dots, x_d) \in X_n\right\}. \tag{5.1}$$

Similarly to Section 4.1, a limit theorem for $a_n^{-1} \circ X_n$ can be obtained. Define a subset $\mathfrak{a}_n(K) \subset \mathbb{R}_+^d$ as

$$\mathfrak{a}_n(K) = \left\{a = (a_1, \dots, a_d) \in \mathbb{R}_+^d: T(a \circ K) \geq 1/n\right\} \tag{5.2}$$

(cf. (1.2)), where

$$a \circ K = \{(a_1 x_1, \dots, a_d x_d): (x_1, \dots, x_d) \in K\}.$$

Furthermore, let

$$\tau_K(x) = T(a \circ K),\ x \in \mathbb{R}_+^d, \tag{5.3}$$

$$\mathcal{T} = \left\{K \in \mathcal{K}: \liminf_{x \to \infty} T(xK) = 0 \text{ for every } x \in \mathbb{R}_+^d\right\},$$

$$q_n(K) = \sup\{t \geq 0: t a_n \in \mathfrak{a}_n(K)\}. \tag{5.4}$$

Note that $q_n(K)$ is the analog of $a_n(K)/a_n$ from Theorem 1.1.

Theorem 5.1 *Assume that for any K from $\mathcal{T}$ there exists the limit of $q_n(K)$ (which is not necessary finite) as $n \to \infty$, and let $\tau_K(x)$ be a regularly varying multivariate function from the class Π_2 having a negative index α. If $a_n \|a_n\|^{-1}$ tends to a certain vector belonging to $\mathrm{Int}\mathbb{R}_+^d$, then $a_n^{-1} \circ X_n$ converges weakly to the random closed set X. The capacity functional of X is equal to*

$$T(K) = \begin{cases} 1 - \exp\{-\lim(q_n(K))^\alpha\} & , \quad K \in \mathcal{T} \\ 1 & , \quad \text{otherwise} \end{cases}$$

The limiting random set X is stable in the following sense. For any $n \geq 1$ and iid copies $X_1, \dots, X_n$ of X there exists $a_n \in \mathbb{R}_+^d$ such that

$$a_n \circ X \overset{d}{\sim} X_1 \cup \dots \cup X_n.$$

It should be pointed out that all results on the pointwise convergence of capacity functionals remain true for unions of random sets in Banach spaces. However, they do not imply the weak convergence, since a random set distribution in Banach space is no longer determined by the corresponding capacity functional.

Enlist several open problems worth mentioning. It has been stated above that the necessity conditions from Section 4.2 do not coincide with the sufficient ones. The problem is to derive *necessary and sufficient* conditions for convergence of unions. It follows from Example 2.2 that the regular variation condition for the function $T_K(x)$ is too restrictive. But in general it cannot be weakened, since for the random set $A = (-\infty, \xi]$ the regular variation condition is necessary and sufficient. On the other hand, Example 2.2 is too artificial, since, in fact, the limiting random set is degenerated on $[0, \infty)$. Maybe, the regular variation condition in Theorem 1.1 is necessary and sufficient in case the limiting random set is non-degenerated in a certain sense.

Multivalued homogeneous functions appear naturally in the scheme of Theorem 4.5. It will be shown in Chapter 6 that some results of Section*4.4 remain true for so-called *multivalued regularly varying functions.*

It seems interesting to prove analogs of limit theorems for *more general norming* schemes than purely multiplicative (e.g., $a_n^{-1}X_n + b_n$ or H_nX_n, where H_n is the sequence of linear operators). Similarly to characterization problems the main obstacle here is the lack of an analog of the Khinchin lemma for random sets distributions. For convex hulls this problem is more simple and can be reduced to limit theorems for coordinate-wise maximums of random vectors.

For *non-identical* distributed summands the limiting distribution corresponds to a certain union-infinitely-divisible random set, see Norberg (1986a). Then the null-array of random sets have to satisfy usual uniform conditions found by Norberg (1986b). However it is rather difficult to reformulate these conditions in terms of regular variation properties of capacities or other analytical properties of random sets distributions. Certainly, some limit theorems for unions of random sets can be derived from general limit theorems for lattice-valued random elements, see Gerritse (1990). Nevertheless, the main problem is to verify the general conditions for particular examples of random closed sets.

It is important to extend the results of this chapter for *weakly dependent* random sets. The corresponding limit theorems can be applied to the study of processes of random growth.

Pancheva (1985,1988) and Zolotarev (1986) considered very general norming scheme for maxima of random variables. It was shown that the use of *non-linear normalizations* allowed to unify max-stable and self-decomposable laws and even to drop the condition of the uniform smallness of summands. Of course, it is interesting to generalizes their approach for random closed sets. The main obstacle here lies in the solving some functional equations on the space of closed sets.

Chapter 5

Almost Sure Convergence of Unions of Random Closed Sets

5.1 Almost Sure Convergence of Random Closed Sets.

Many investigations concern with finding the almost sure limit for a random sample in $\mathbb{R}^d$ or its convex hull as the sample size increases. This problem was solved in Davis, Mulrow and Resnick (1988), where further references and commentaries can be found.

This chapter is intended to prove a *strong law of large numbers for unions* of random sets. It deals with the almost sure convergence of normalized unions to a non-random limit. Note that the results for samples of random vectors are imbedded in our scheme since a random vector is a single-point random set.

Let A be a random closed set in the Euclidean space $\mathbb{R}^d$, and let $A_1, \ldots, A_n$ be iid copies of A. Denote

$$Y_n = a_n^{-1}(A_1 \cup \cdots \cup A_n). \tag{1.1}$$

Of course, Y_n is a random closed set. We shall find conditions which ensure convergence and the limit of the sequence Y_n as $n \to \infty$.

The convergence of closed sets in $\mathcal{F}$ ($\mathcal{F}$-convergence) was defined in Section 1.1, see also Matheron (1975). Recall that a sequence F_n, $n \geq 1$, of closed sets is said to converge to F if the following conditions are satisfied:

(F1) If $K \cap F = \emptyset$ for a certain compact K from $\mathcal{K}$, then there exists a number $N > 0$ such that $K \cap F_n = \emptyset$ whenever $n \geq N$.

(F2) If $G \cap F \neq \emptyset$ for a certain G from the class $\mathcal{G}$ of all open sets, then there exists a number $N > 0$ such that $G \cap F_n \neq \emptyset$ whenever $n \geq N$.

In our scheme the sets Y_n, $n \geq 1$, are allowed to be unbounded, so that we cannot use convergence in $\mathcal{K}$ (compare with Davis et al. (1988), where the convergence in $\mathcal{K}$ was investigated).

A random set Y_n is said to converge to Y almost surely if $Y = \mathcal{F}-\lim Y_n$ with probability one. We then write $Y_n \xrightarrow{\mathcal{F}} Y$ a.s as $n \to \infty$. The almost sure convergence in $\mathcal{K}$ is defined similarly.

The conditions **(F1)** and **(F2)** can be safely reformulated for the sets K and G belonging to some countable subfamilies of $\mathcal{K}$ and $\mathcal{G}$ respectively. Then we can render these conditions for a sequence of random sets with a non-random limit, see Davis et al. (1988).

Lemma 1.1 (Davis, Mulrow and Resnick) *Let Y_n, $n \geq 1$, be a sequence of random closed sets and let Y be a non-random closed set. Then $Y_n \xrightarrow{\mathcal{F}} Y$ almost surely as $n \to \infty$ iff the following conditions are valid:*

(R1) *If $K \cap Y = \emptyset$ for a certain K from $\mathcal{K}$, then*

$$\mathbf{P}\{Y_n \cap K \neq \emptyset \text{ i.o.}\} = \mathbf{P}\left\{\bigcap_{n=1}^{\infty} \bigcup_{m=n}^{\infty} \{Y_n \cap K \neq \emptyset\}\right\} = 0.$$

(R2) *If $G \cap Y \neq \emptyset$ for a certain G from $\mathcal{G}$ then*

$$\mathbf{P}\{Y_n \cap G = \emptyset \text{ i.o.}\} = \mathbf{P}\left\{\bigcap_{n=1}^{\infty} \bigcup_{m=n}^{\infty} \{Y_n \cap G = \emptyset\}\right\} = 0.$$

These conditions may be weakened by replacing the class $\mathcal{K}$ in **(R1)** and $\mathcal{G}$ in **(R2)** with some their sub-classes $\mathcal{M}$ and $\mathcal{M}'$ respectively. It is rather easy to show that we can choose the class of all balls (open balls) instead of $\mathcal{M}$ (respectively $\mathcal{M}'$). We may as well take parallelepipeds as their elements.

The classes $\mathcal{M}$ and $\mathcal{M}'$ are said to *determine $\mathcal{F}$-convergence* if Lemma 1.1 is valid after replacing $\mathcal{K}$ with $\mathcal{M}$ and $\mathcal{G}$ with $\mathcal{M}'$.

We always assume that the following assumptions is valid.

ASSUMPTIONS.

1. Each K from $\mathcal{M}$ coincides with the closure of its interior (i.e. K is canonically closed).

2. The interior $\text{Int}K$ is the limit of an increasing sequence of sets from $\mathcal{M}$.

3. The class $\mathcal{M}'$ consists of interiors of all sets from $\mathcal{M}$, i.e.

$$\mathcal{M}' \supseteq \{\text{Int}K : K \in \mathcal{M}\}.$$

4. For any $c > 0$

$$c\mathcal{M} = \{cK : K \in \mathcal{M}\} = \mathcal{M}.$$

Let $\mathbb{S}^{d-1}$ be the unit sphere in $\mathbb{R}^d$. Assume that $\mathbb{S}^{d-1}$ is furnished with the topology induced by the standard topology in $\mathbb{R}^d$. Then the class

$$\mathcal{M} = \left\{\{ux : u \in S, a \leq x \leq b\} : S \text{ is a closed subset of } \mathbb{S}^{d-1}, 0 \leq a < b\right\} \quad (1.2)$$

can be used in Lemma 1.1 instead of $\mathcal{K}$ as well. Indeed, each K from $\mathcal{K}$, such that $K \cap F = \emptyset$, can be covered with a collection of sets $K_1, \ldots, K_n$ belonging to $\mathcal{M}$ and missing F. Besides, any open G hitting F contains a set $G_1 \in \mathcal{M}'$ which hits F too. It can be also shown that the set S in (1.2) can be assumed to take values only in the class of canonically closed subsets of $\mathbb{S}^{d-1}$.

5.2 Regularly Varying Capacities.

Investigating random closed sets, we deal with *capacities* instead of functions. Now we translate to capacities some notions from multivariate regular variation theory, see Section 1.6.

Let $\mathcal{M}$ be a sub-class of $\mathcal{F}$ and let $R: \mathcal{M} \longrightarrow [0,\infty]$ be a nonnegative capacity. Suppose that R is an upper semicontinuous decreasing capacity without any restrictions on signs of higher differences inherent to Choquet capacities, see Section 1.2 and Matheron (1975).

The capacity R is said to be *regularly varying* on $\mathcal{M}$ with the limit capacity Λ if, for all F from $\mathcal{M}$,

$$\lim_{t\to\infty} \frac{R(tF)}{g(t)} = \Lambda(F), \tag{2.1}$$

where $g: (0,\infty) \longrightarrow (0,\infty)$ is a regularly varying function of index β, see Seneta (1976) and Section 1.6. We then write $R \in \mathrm{RV}(\beta, \mathcal{M}, \Lambda, g)$.

The limiting capacity $\Lambda(F)$ is allowed to take zero or infinite values. However we suppose that Λ is not equal to zero or infinity identically.

It is easy to prove that Λ is a decreasing functional on $\mathcal{M}$, and, for any $C > 0$, F_1 from $\mathcal{M}$, the limit (2.1) exists for the set $F = CF_1$. Moreover,

$$\Lambda(CF) = C^{\beta}\Lambda(F_1).$$

Lemma 2.1 *Let T be the capacity functional of a certain random closed set A, and let*

$$R(K) = -\log T(K) \tag{2.2}$$

belong to $\mathrm{RV}(\beta, \mathcal{M}, \Lambda, g)$ with positive β. Then, for any F_1 and F_2 from $\mathcal{M}$, the limit (2.1) exists for the set $F = F_1 \cup F_2$, and also

$$\Lambda(F) = \min(\Lambda(F_1), \Lambda(F_2)). \tag{2.3}$$

PROOF. It is evident that for $i = 1, 2$

$$\begin{aligned}\limsup_{t\to\infty} \frac{R(tF)}{g(t)} &\le \lim_{t\to\infty} \frac{R(tF_i)}{g(t)} \\ &\le \min(\Lambda(F_1), \Lambda(F_2)).\end{aligned}$$

If either $\Lambda(F_1)$ or $\Lambda(F_2)$ is equal to zero then (2.3) is evident. Let both $\Lambda(F_1)$ and $\Lambda(F_2)$ be finite and non-vanishing. Then, for any $\varepsilon > 0$ and sufficiently large t,

$$T(tF_i) \le \exp\{-g(t)\Lambda(F_i)(1-\varepsilon)\}, i = 1, 2. \tag{2.4}$$

Hence, by subadditivity of T, we get

$$\begin{aligned}\liminf_{t\to\infty} \frac{R(tF)}{g(t)} &\ge \liminf_{t\to\infty} -\frac{\log(T(tF_1) + T(tF_2))}{g(t)} \\ &\ge \liminf_{t\to\infty} -\frac{\log(2\exp\{-g(t)\min(\Lambda(F_1), \Lambda(F_2))(1-\varepsilon)\})}{g(t)} \\ &= \min(\Lambda(F_1), \Lambda(F_2))(1-\varepsilon).\end{aligned}$$

Note that $g(t)^{-1} \to 0$ as $t \to \infty$, since $\beta > 0$. Hence (2.3) is valid.

If $\Lambda(F_1) = \infty$, then (2.4) is replaced with the inequality

$$T(tF_1) \leq \exp\{-g(t)C\},$$

which holds for any positive C and sufficiently large t. Then, for $C > \Lambda(F_2)$,

$$\begin{aligned}
\liminf_{t\to\infty} \frac{R(tF)}{g(t)} &\geq \liminf_{t\to\infty} -\frac{\log(\exp\{-g(t)C\} + \exp\{-g(t)\Lambda(F_2)(1-\varepsilon)\})}{g(t)} \\
&\geq \liminf_{t\to\infty} -\frac{\log(2\exp\{-g(t)\Lambda(F_2)(1-\varepsilon)\})}{g(t)} \\
&= \Lambda(F_2)(1-\varepsilon) \\
&= \min(\Lambda(F_1), \Lambda(F_2))(1-\varepsilon). \quad \square
\end{aligned}$$

Below we always associate the capacity R with the capacity functional of a certain random closed set A by means of (2.2). It follows from Lemma 2.1 that Λ is a *minitive capacity* (compare with maxitive capacities introduced in Norberg, 1986b). Nevertheless, the value $\Lambda(F)$ cannot be represented as the minimum value of $\Lambda(\{x\})$ for x belonging to F, since, in general, the class $\mathcal{M}$ does not contain single-point sets and $\Lambda(\{x\})$ can be infinite.

The functional Λ is said to be *strictly monotone* (decreasing) on $\mathcal{M}$ if $\Lambda(K_1) > \Lambda(K)$ for any K, K_1 from $\mathcal{M}$ such that $K_1 \subset \text{Int}K$, $\Lambda(K) < \infty$.

Denote for any compact K

$$\hat{K} = \cup\{sK : s \geq 1\}.$$

It is evident that $\hat{K}$ is closed and $s\hat{K} \subseteq \hat{K}$ for all $s \geq 1$.

Lemma 2.2 *Suppose that $R \in \text{RV}(\beta, \mathcal{M}, \Lambda, g)$, $\beta > 0$, and the capacity Λ is continuous in the following sense: $\Lambda(K_n) \downarrow \Lambda(F)$ as long as $K_n \uparrow F$, whatever F from $\mathcal{F}$ and a sequence of compacts K_n, $n \geq 1$, belonging to $\mathcal{M}$ may be. Moreover, let the limit (2.1) exist for the limiting set F. Then the limit (2.1) exists for the set $F = \hat{K}$ and also $\Lambda(\hat{K}) = \Lambda(K)$, whatever K from $\mathcal{M}$ may be.*

PROOF. It is obvious that $K_n \uparrow \hat{K}$ for K_n defined as

$$K_n = \bigcup_{i=1}^{n} s_{in} K,$$

where $\{s_{1n}, \ldots, s_{nn}\} \subset [1, \infty)$ for all n. The homogeneous property of Λ yields

$$\Lambda(s_{in}K) = s_{in}^{\beta}\Lambda(K) \geq \Lambda(K).$$

Then, by Lemma 2.1, we get

$$\Lambda(K_n) = \min_{1\leq i\leq n} \Lambda(s_{in}K) \geq \Lambda(K).$$

It follows from the condition of Lemma 2.2 that

$$\Lambda(\hat{K}) = \lim \Lambda(K_n) \geq \Lambda(K).$$

On the other hand, $\Lambda(K) \leq \Lambda(\hat{K})$, since $\hat{K} \supseteq K$. Thus, $\Lambda(\hat{K}) = \Lambda(K)$. □

Note that the condition of Lemma 2.2 is valid in case (2.1) is satisfied uniformly on $\mathcal{K}$. Namely,

$$\lim_{t\to\infty} \frac{R(tK_t)}{g(t)} = \Lambda(K)$$

as soon as $K_t \uparrow K \in \mathcal{K}$ as $t \to \infty$, cf. the class Π_2 in Section 1.6.

For any functional $\Lambda: \mathcal{M} \longrightarrow [0, \infty]$ denote

$$Z(\Lambda; \mathcal{M}) = \left(\bigcup \{\mathrm{Int}F:\ F \in \mathcal{M}, \Lambda(F) > 1\}\right)^c. \tag{2.5}$$

Lemma 2.3 *Let Λ be a limiting capacity in (2.1). Then*

$$sZ(\Lambda; \mathcal{M}) \subset Z(\Lambda; \mathcal{M}),\ s \geq 1.$$

PROOF immediately follows from the inequality

$$\Lambda(sF) = s^\beta \Lambda(F) > \Lambda(F)$$

for all $s > 1$. □

5.3 A Strong Law of Large Numbers for Unions of Random Closed Sets.

Now that we have introduced all necessary notions we investigate almost sure convergence of the random set Y_n defined in (1.1), as $a_n \to \infty$, $n \to \infty$. The following theorem (see Molchanov, 1993c) resembles to some extent Theorem 2.1 from the cited work by Davis et al. (1988) which, in fact, dealt with the similar problem for single-point random sets $A_i = \{\xi_i\}$.

Theorem 3.1 *Let A be a random closed set with the capacity functional T, and let the class $\mathcal{M}$ determine $\mathcal{F}$-convergence. Define the capacity R with possibly infinite values by (2.2). Furthermore, let $R \in \mathrm{RV}(\beta, \mathcal{M}, \Lambda, g)$ for $\beta > 0$. Suppose that Λ is a strictly monotone capacity on $\mathcal{M}$ and for any K from $\mathcal{M}$*

$$\lim_{t\to\infty} \frac{R(t\hat{K})}{g(t)} = \Lambda(\hat{K}) = \Lambda(K). \tag{3.1}$$

Since $\beta > 0$, we can define a_n to satisfy $g(a_n) \sim \log n$. Then

$$Y_n = a_n^{-1}(A_1 \cup \cdots \cup A_n) \xrightarrow{\mathcal{F}} Z(\Lambda; \mathcal{M}) \text{ a.s. as } n \to \infty, \tag{3.2}$$

and

$$\mathrm{conv}(Y_n) \xrightarrow{\mathcal{F}} \mathrm{conv}(Z(\Lambda; \mathcal{M})) \text{ a.s. as } n \to \infty.$$

PROOF. Verify the conditions of Lemma 1.1 for the class $\mathcal{M}$ and the class $\mathcal{M}' = \{\text{Int}K: K \in \mathcal{M}\}$.

Let K belong to $\mathcal{M}$ and miss $Z(\Lambda; \mathcal{M})$. Then

$$K \subset \bigcup \{\text{Int}F: \Lambda(F) > 1, F \in \mathcal{M}\}.$$

Hence K is covered by the finite collection of sets $\text{Int}F_i$, $1 \le i \le m$, for F_i belonging to $\mathcal{M}$ such that $\Lambda(F_i) > 1$.

It follows from (2.1) and the choice of a_n that

$$\lim_{n\to\infty} \frac{R(a_n F_i)}{\log n} = \Lambda(F_i).$$

Lemma 2.1 yields

$$\begin{aligned}\lim_{n\to\infty} \frac{R(a_n K)}{\log n} &= \Lambda(K) \\ &\ge \min_{1\le i\le m} \Lambda(F_i) \\ &= a > 1.\end{aligned}$$

It follows from (3.1) that

$$\lim_{n\to\infty} \frac{R(a_n \hat{K})}{\log n} = \Lambda(\hat{K}) = \Lambda(K) > a.$$

Pick $\zeta > 0$ such that $a - \zeta > 1$. Then, for all sufficiently large n,

$$T(a_n \hat{K}) \le n^{-(a-\zeta)}. \tag{3.3}$$

Note that

$$\mathbf{P}\left\{Y \cap \hat{K} \neq \emptyset \text{ i.o.}\right\} = \mathbf{P}\left\{A_{i_n} \cap a_n \hat{K} \neq \emptyset \text{ i.o.}\right\},$$

where $1 \le i_n \le n$, $n \ge 1$. It is easy to show that the sequence i_n, $n \ge 1$, is unbounded. Since $a_n \hat{K} \subset a_{n+1} \hat{K}$ for $n \ge 1$, we get

$$\begin{aligned}\mathbf{P}\left\{Y_n \cap \hat{K} \neq \emptyset \text{ i.o.}\right\} &= \mathbf{P}\left\{(A_1 \cup \cdots \cup A_n) \cap a_n \hat{K} \neq \emptyset \text{ i.o.}\right\} \\ &= \mathbf{P}\left\{A_{i_n} \cap a_n \hat{K} \neq \emptyset \text{ i.o.}\right\} \\ &\le \mathbf{P}\left\{A_n \cap a_n \hat{K} \neq \emptyset \text{ i.o.}\right\} \\ &\le \mathbf{P}\left(\bigcap_{n=1}^{\infty} \bigcup_{m=n}^{\infty} \{A_n \cap a_n \hat{K} \neq \emptyset\}\right).\end{aligned}$$

The latter probability is equal to zero due to (3.3) and the Borel-Cantelli lemma, since

$$\begin{aligned}\sum_{n=1}^{\infty} \mathbf{P}\left\{A_n \cap a_n \hat{K} \neq \emptyset\right\} &= \sum_{n=1}^{\infty} T(a_n \hat{K}) \\ &\le \sum_{n=1}^{\infty} n^{-(a-\zeta)} < \infty.\end{aligned}$$

Let x belong to $G \cap Z(\Lambda; \mathcal{M})$ for a certain G from $\mathcal{M}'$. Then

$$G \not\subset \bigcup \{\mathrm{Int}K\colon \Lambda(K) > 1, K \in \mathcal{M}\}.$$

Choose an open neighborhood $U(x) \subset G$ and pick K and K_1 from $\mathcal{M}$ such that

$$U(x) \subset K_1 \subset \mathrm{Int}K \subset K \subset G.$$

If $\Lambda(K) \geq 1$, then $\Lambda(K_1) > 1$, since Λ is strictly monotone. Hence

$$x \in \mathrm{Int}K_1 \text{ and } \Lambda(K_1) > 1,$$

so that $x \notin Z(\Lambda; \mathcal{M})$. Thus, $\Lambda(K) = a < 1$ and $K \cap Z(\Lambda; \mathcal{M}) \neq \emptyset$. Clearly,

$$\mathbf{P}\{Y_n \cap G = \emptyset \text{ i.o. }\} \leq \mathbf{P}\{Y_n \cap K = \emptyset \text{ i.o. }\}.$$

Pick $\zeta > 0$ such that $a + \zeta < 1$. Then

$$T(a_n K) \geq n^{-(a+\zeta)}$$

for all sufficiently large n. Thus

$$\begin{aligned}\mathbf{P}\{(A_1 \cup \cdots \cup A_n) \cap a_n K = \emptyset\} &= (1 - T(a_n K))^n \\ &\leq \exp\{-nT(a_n K)\} \\ &\leq \exp\{nn^{-(a+\zeta)}\} \\ &= \exp\{-n^{1-(a+\zeta)}\}.\end{aligned}$$

Since $\delta = 1 - (a + \zeta) > 0$, we get

$$\sum_{n=1}^{\infty} \mathbf{P}\{Y_n \cap K = \emptyset\} \leq \sum_{n=1}^{\infty} \exp\{-n^{\delta}\} < \infty.$$

Hence $\mathbf{P}\{Y_n \cap G = \emptyset \text{ i.o. }\} = 0$.

Thus, both conditions of Lemma 1.1 are valid. The convergence of convex hulls follows from the continuity of the function $F \longrightarrow \mathrm{conv}(F)$ with respect to $\mathcal{F}$-convergence, see Matheron (1975). □

Corollary 3.2 *Suppose that the conditions of Theorem 3.1 are valid, and* $\Lambda(K_0^c) \neq 0, \infty$ *for a certain convex compact* K_0 *such that* $0 \in \mathrm{Int}K_0$, $K_0^c \in \mathcal{M}'$. *Then* Y_n *almost surely converges to* $Z(\Lambda; \mathcal{M})$ *with respect to the Hausdorff metric.*

PROOF. The convergence in the Hausdorff metric is equivalent to $\mathcal{K}$-convergence of compacts (see Section1.1). We have to check additionally that

$$\mathbf{P}\left\{\sup_{n\geq 1} \sup_{x \in Y_n} \|x\| < \infty\right\} = 1,$$

see also Davis et al. (1988). It suffices to show that

$$\mathbf{P}\left\{\sup_{n\geq 1}\inf\{t>0\colon\ Y_n\subset tK_0\}<\infty\right\}=$$
$$=\mathbf{P}\left\{\sup_{n\geq 1}a_n^{-1}\inf\{t>0\colon\ A_1\cup\cdots\cup A_n\subset tK_0\}<\infty\right\}=1.$$

Denote

$$\zeta_n=\inf\{t>0\colon\ A_1\cup\cdots\cup A_n\subset tK_0\}.$$

Then $\zeta_n=\max(\eta_1,\ldots,\eta_n)$ for iid random variables $\eta_1,\ldots,\eta_n$ with the common distribution

$$\mathbf{P}\{\eta_1>y\}=\mathbf{P}\{A\cap yK_0^c\neq\emptyset\}.$$

Hence $-\log(\mathbf{P}\{\eta_1>y\})$ is a regularly varying function. This is sufficient for almost sure stability of $\sup_{n\geq 1}a_n^{-1}\zeta_n$, see Resnick and Tomkins (1973). □

It can be shown that for a *single-point set* $A=\{\xi\}$ all conditions of Theorem 3.1 follow from the conditions on the distribution of ξ imposed in Theorem 2.1 of Davis et al. (1988). In this case it is reasonable to choose the class of all parallelepipeds as the class $\mathcal{M}$. Note that in this case the functional Λ of any parallelepiped K depends on the lower-left vertex of K only.

Preserve all notations from Davis et al. (1988). Let $A=\{\xi\}$, and let ξ be distributed in $\mathbb{R}_+^d=[0,\infty)^d$ only. Put

$$r(x)=-\log\mathbf{P}\{\xi\leq x\},$$

where the inequality is understood coordinate-wisely. Choose parallelepipeds to be elements of $\mathcal{M}$.

Theorem 3.3 (Davis, Mulrow, Resnick) *Let $r(x)$ be a regularly varying function on $\mathbb{R}_+^d\setminus\{0\}$, with the index of variation $\beta>0$, i.e.*

$$\lim_{t\to\infty}\frac{r(tx)}{g(t)}=\Lambda(x),$$

where

$$\Lambda(tx)=t^\beta\Lambda(x),\quad t>0,\ x\in\mathbb{R}_+^d\setminus\{0\},$$

and g is a regularly varying function of index $\beta>0$. Suppose that λ is strictly monotone (increasing) coordinate-wisely. Then in $\mathcal{K}$

$$a_n^{-1}\{\xi_1,\ldots,\xi_n\}\to\{x\in\mathbb{R}_+^d\colon\ \lambda(x)\leq 1\}\ \ a.s.\ as\ n\to\infty.$$

PROOF. Check the conditions of Theorem 3.1. Introduce the capacity $R(K)$ by

$$R(K)=-\log\mathbf{P}\{\xi\in K\}.$$

It suffices to prove that $R\in\mathrm{RV}(\beta,\mathcal{M},\Lambda,g)$, and $\Lambda(K)$ for any parallelepiped K is equal to $\lambda(a)$, where a is its lower-left vertex. Indeed, then $\Lambda(\check{K})=\Lambda(K)$, Λ is strictly monotone on $\mathcal{M}$, so that the statement of Theorem follows from Theorem 3.1.

So let K be a parallelepiped $[a,b]=[a_1,b_1]\times\cdots\times[a_d,b_d]$. Denote $F(x)=\mathbf{P}\{\xi\ge x\}$ for $x\in\mathbb{R}_+^d$ (all inequalities are coordinate-wise). Then, similarly to Davis et al. (1988),

$$\begin{aligned}\mathbf{P}\{\xi\in tK\} &= F(ta)-\sum_{i=1}^{d}F(tx(i))\\ &+\sum_{1\le i<j\le d}F(tx(i,j))-\cdots+(-1)^dF(tb),\end{aligned}$$

where $x(i)$ is the vector with the ith component b_i and whose p^{th} component is a_p, $p\neq i$, etc. It follows from the assumptions of Theorem that for any $\zeta>0$ there exists $t_0=t_0(\zeta,a,b)$ such that

$$\exp\{-g(t)(\lambda(z)+\zeta)\}\le F(tz)\le\exp\{-g(t)(\lambda(z)-\zeta)\},$$

where $z=a$, or b, or $x(i),x(i,j)$ etc. for some $i,j,\ldots$.

Then

$$\begin{aligned}\mathbf{P}\{\xi\in tK\} &\ge q(t)^{-(\lambda(a)+\zeta)}-\sum_{i=1}^{d}q(t)^{-(\lambda(x(i))-\zeta)}\\ &+\sum_{1\le i<j\le d}q(t)^{-(\lambda(x(i,j))+\zeta)}-\cdots+(-1)^dq(t)^{-(\lambda(b)+(-1)^d\zeta)},\end{aligned}$$

where $q(t)=\exp\{g(t)\}$. Hence

$$\begin{aligned}\mathbf{P}\{\xi\in tK\} &\ge q(t)^{-(\lambda(a)+\zeta)}\Bigg[1-\sum_{i=1}^{d}q(t)^{-(\lambda(x(i))-\lambda(a)-2\zeta)}\\ &+\sum_{1\le i<j\le d}q(t)^{-(\lambda(x(i,j))-\lambda(a))}-\cdots+(-1)^dq(t)^{-(\lambda(b)-\Lambda(a)-\zeta+(-1)^d\zeta)}\Bigg].\end{aligned}$$

Since λ is increasing, letting ζ be sufficiently small yields

$$\begin{aligned}\mathbf{P}\{\xi\in tK\} &\ge q(t)^{-(\lambda(a)+\zeta)}\Bigg[1-\sum_{i=1}^{d}q(t)^{-c(i)}\\ &+\sum_{1\le i<j\le d}q(t)^{-c(i,j)}-\cdots+(-1)^dq(t)^{-c}\Bigg],\end{aligned}$$

where $c(i)>0$, $c(i,j)>0,\ldots,c>0$. It was established in Davis et al. (1988) that the term in brackets is no less that $1/2$ for all sufficiently large t. Thus, for $K=[a,b]$,

$$T(tK)=\mathbf{P}\{\xi\in tK\}\ge\frac{1}{2}\exp\{-g(t)(\lambda(a)+\zeta)\}.$$

Hence, for a certain t_0 and all $t\ge t$,

$$\begin{aligned}\frac{1}{2}\exp\{-g(t)(\lambda(a)+\zeta)\}\le T(tK) &\le \mathbf{P}\{\xi\ge ta\}\\ &= F(at)\le\exp\{-g(t)(\lambda(a)-\zeta)\}.\end{aligned}$$

Thus

$$\lambda(a) - \zeta \le \frac{R(tK)}{g(t)} \le \lambda(a) + \zeta - \frac{\log 2}{g(t)}.$$

Therefore

$$\lim_{t\to\infty} \frac{R(tK)}{g(t)} = \Lambda(K) = \Lambda([a,b]) = \lambda(a).$$

Thus, in complete accordance with Davis et al. (1988), $a_n^{-1}\{\xi_1, \ldots, \xi_n\}$ converges in $\mathcal{K}$ to the set

$$Z(\Lambda; \mathcal{M}) = \left(\bigcup\{[x,\infty)\colon \lambda(x) > 1\}\right)^c = \left\{x \in \mathbb{R}_+^d\colon \lambda(x) \le 1\right\}. \quad \square$$

It should be noted also that the *lack of preferable directions* in Theorem 3.1 (in contrast to Theorem 3.3 where ξ is distributed within $\mathbb{R}_+^d$ and the distribution function F is defined on upper-right unbounded sets only) makes it possible to apply it for random samples in all quadrants of $\mathbb{R}^d$ without any changes, cf Davis et al. (1988).

We can also apply Theorem 3.1 to the random set

$$A = (-\infty, \xi] = (-\infty, \xi_1] \times \cdots \times (-\infty, \xi_d].$$

Then for a parallelepiped $[a,b] = [a_1, b_1] \times \cdots \times [a_d, b_d]$

$$\mathbf{P}\{A \cap t[a,b] \neq \emptyset\} = \mathbf{P}\{\xi \ge ta\} = F(ta).$$

Thus, regular variation of the function $-\log F(ta)$ for any $a \in \mathbb{R}_+^d$ with the strictly monotone limiting function λ ensures almost sure convergence of the appropriate normalized unions to the deterministic limit

$$\left\{y \in \mathbb{R}^d\colon\ y \le x \in \mathbb{R}_+^d, \lambda(x) \le 1\right\}.$$

In case we are interested in the convergence of *convex hulls* only, a simpler condition can be obtained.

Let $Y_n = a_n^{-1}\mathrm{conv}(A_1 \cup \cdots \cup A_n)$, where $A_1, \ldots, A_n$ are iid copies of a compact random set A. Then, in terms of support functions,

$$s_{Y_n}(u) = a_n^{-1} \max\{s_{A_1}(u), \ldots, s_{A_n}(u)\}, u \in \mathbb{S}^{d-1}.$$

The following theorem simply follows from Resnick and Tomkins (1973).

Theorem 3.4 *Let for any $u \in \mathbb{S}^{d-1}$ and $x > 0$ there exists the finite limit*

$$\lim_{t\to\infty} \frac{-\log \mathbf{P}\{s_A(ux) \ge t\}}{g(t)} = \lambda(ux).$$

Then Y_n a.s. converges in $\mathcal{K}$ to the set

$$\left\{ux\colon\ u \in \mathbb{S}^{d-1}, x \ge 0, \lambda(ux) \ge 1\right\}.$$

5.4 Almost Sure Limits for Unions of Special Random Sets.

Now that a general theorem on almost sure convergence of scaled unions has been derived, we consider one special but rather general example of random sets and the corresponding law of large numbers.

Let h be a regularly varying function on $\mathbb{R}^m$ with the limiting function λ (see Section 1.6), and let $\operatorname{ind} h = \alpha > 0$, i.e.

$$\lambda(tx) = t^{\alpha}\lambda(x)$$

for all $t > 0$, $x \neq 0$. Suppose that λ is non-vanishing on $\mathbb{S}^{m-1}$. Moreover, let $h \in \Pi_2$, i.e. h satisfy the *uniformity condition* (see Yakimiv, 1981, de Haan and Resnick, 1987):

$$\lim_{t\to\infty} \sup_{\|x\|=1} \left|\frac{h(tx)}{h(te)} - \lambda(x)\right| = 0 \tag{4.1}$$

for a certain e from $\mathbb{R}^d \setminus \{0\}$, see also Section 1.6. Sometimes it is convenient to put $e = \mathbb{I} = (1, \ldots, 1)$. It was proven in Yakimiv (1981) that λ is a continuous function.

Let ξ be a random vector in $\mathbb{R}^m$ with the probability density $\exp\{-h(x)\}$, and let $M: \mathbb{R}^m \longrightarrow \mathcal{K}$ be a continuous in the Hausdorff metric multivalued homogeneous function of index $\eta > 0$, i.e.

$$M(tu) = t^{\eta}M(u)$$

for all $t > 0$, $u \neq 0$. Suppose also that $M(u) = \{0\}$ iff $u = 0$. It is easy to show that $\|M(e)\|$ is continuous on $\mathbb{S}^{m-1}$. Hence

$$\sup_{\|x\|=1} \|M(x)\| < \infty.$$

Note that the values of M are compact subsets of $\mathbb{R}^d$, so m is allowed to be different from d.

Denote

$$\begin{aligned} A &= M(\xi), \\ g(t) &= h(t^{1/\eta}e). \end{aligned}$$

Let $A_1, A_2, \ldots$ be independent copies of the RACS A.

For simplicity suppose that for any open cone $\Gamma \subset \mathbb{R}^d$

$$\mu_{m-1}\left(\{e \in \mathbb{S}^{m-1}\colon M(e) \cap \Gamma \neq \emptyset\}\right) > 0 \tag{4.2}$$

where μ_{m-1} is the $(m-1)$-dimensional Lebesgue measure on $\mathbb{S}^{m-1}$. Otherwise the range of possible values of A is a certain cone $\mathbb{G}$, so that all results below can be safely reformulated after replacing $\mathbb{R}^d$ with $\mathbb{G}$ and $\mathbb{S}^{d-1}$ with $\mathbb{S}^{d-1} \cap \mathbb{G}$.

For each $S \subset \mathbb{S}^{d-1}$ denote

$$\begin{aligned} \Gamma_S &= \{xv\colon x \geq 0, v \in S\}, \\ L(S) &= \{e \in \mathbb{S}^{m-1}\colon M(e) \cap \Gamma_S \neq \emptyset\}, \\ q_S(e) &= \|M(e) \cap \Gamma_S\|, e \in L(S). \end{aligned}$$

For a single-point set $S = \{v\}$ the corresponding notations are replaced with Γ_v, $L(v)$ and $q_v(e)$ respectively.

Theorem 4.1 *Suppose that the following assumption is valid*

(i) *for any canonically closed set $S \subset \mathbb{S}^{d-1}$, $e_0 \in L(S)$, $\eta > 0$ and u_0 from $M(e_0) \cap \Gamma_S$ there exists $e_1 \in L(S)$ such that*

$$M(e_1) \cap \mathrm{Int}\Gamma_S \cap B_\eta(u_0) \neq \emptyset.$$

Pick a_n such that $g(a_n) \sim \log n$. Then, in $\mathcal{K}$,

$$Y_n = a_n^{-1}(A_1 \cup \cdots \cup A_n) \to Z = \left\{vx\colon\, v \in \mathbb{S}^{d-1}, 0 \leq x \leq f(v)\right\} \quad a.s. \text{ as } n \to \infty,$$

where

$$f(v) = \sup\left\{\frac{q_v(e)}{\lambda(e)^{\eta/\alpha}}\colon\, e \in L(v)\right\}.$$

We begin with a lemma.

Lemma 4.2 *Let $F \subset (0, +\infty)$ be a finite union of disjoint segments of positive lengths with possibly infinite right end-points, and let $g(y) = e^{-y}y^{\gamma-1}$, $\gamma > 0$. Then*

$$\int_{Ft} g(y)dy \sim g(t \inf F) \quad as \ t \to \infty. \tag{4.3}$$

PROOF. Let $F = [a, \infty)$. Then the statement of Lemma 4.2 follows from

$$\int_{at}^{\infty} g(y)dy \sim g(at) \text{ as } t \to \infty.$$

It is evident that $g(bt)/g(at) \to 0$ as $t \to \infty$ in case $a < b$. Thus, for $F = [a, b]$ we get

$$\lim_{t \to \infty} \frac{\int_{at}^{bt} g(y)dy}{g(at)} = 1.$$

By induction, (4.3) is valid for any F. □

PROOF OF THEOREM 4.1. For any K from $\mathcal{K}$ denote

$$\mathcal{L}_K = \{u \in \mathbb{R}^m\colon\, M(u) \cap K \neq \emptyset\}.$$

It is easy to show that $\mathcal{L}_{tK} = t^{1/\eta}\mathcal{L}_K$ and $\|x\| \geq \delta$ for some $\delta > 0$ and all x from $\mathcal{L}_K$ in case $0 \notin K$.

It is obvious that the limiting set Z contains the origin. Suppose that $0 \notin K$, i.e. K misses the origin. Then

$$\begin{aligned} T(tK) &= \mathbf{P}\{tK \cap A \neq \emptyset\} \\ &= \mathbf{P}\{\xi \in \mathcal{L}_{tK}\} \\ &= \mathbf{P}\left\{\xi \in t^{1/\eta}\mathcal{L}_K\right\} \\ &= \int_{\mathcal{L}_K} \exp\left\{-h(t^{1/\eta}u)\right\} t^{d/\eta} du. \end{aligned}$$

It follows from (4.1) and Lemma 6.2 from Davis et al. (1988) (see also Lemma 6.3.2) that for all $\varepsilon > 0$ and sufficiently large t

$$T(tK) \leq I_1(t) = \int_{\mathcal{L}_K} \exp\left\{-(1-\varepsilon)\lambda(u)\|u\|^{-\varepsilon} g(t)\right\} t^{d/\eta} du, \tag{4.4}$$

$$T(tK) \geq I_2(t) = \int_{\mathcal{L}_K} \exp\left\{-(1+\varepsilon)\lambda(u)\|u\|^{\varepsilon} g(t)\right\} t^{d/\eta} du. \tag{4.5}$$

It suffices to verify the conditions of Theorem 3.1 for the class $\mathcal{M}$ defined by (1.2). Let

$$K = \{ux \colon u \in S, a \leq x \leq b\}, \tag{4.6}$$

where $0 < a < b < \infty$ and S is a canonically closed subset of $\mathbb{S}^{d-1}$.

Denote additionally for e belonging to $L(S)$

$$\begin{aligned} \mathcal{L}_K(e) &= \{r \geq 0 \colon r^\eta M(e) \cap K \neq \emptyset\}, \\ \ell_K(e) &= \inf \mathcal{L}_K(e) = \left(\frac{a}{q_S(e)}\right)^{1/\eta}. \end{aligned} \tag{4.7}$$

The function $q_S(e)$ is bounded on $L(S)$, since

$$\tilde{q} = \sup\{q_S(e) \colon e \in L(S)\} \leq \sup\{\|M(e)\| \colon e \in \mathbb{S}^{m-1}\} < \infty. \tag{4.8}$$

Thus

$$f_K = \inf\{\ell_K(e) \colon e \in L(S)\} \geq (a/\tilde{q})^{1/\eta}.$$

Since

$$\mathcal{L}_K = \{er \colon e \in L(S), r \in \mathcal{L}_K(e)\},$$

the integral $I_1(t)$ from (4.4) is equal to

$$I_1(t) = \int_{L(S)} \mu_{m-1}(de) \int_{\mathcal{L}_K(e)} \exp\left\{-(1-\varepsilon)\lambda(e) r^{\alpha-\varepsilon} g(t)\right\} t^{d/\eta} r^{d-1} dr.$$

Letting y be equal to $(1-\varepsilon) r^{\alpha\varepsilon} \lambda(e) g(t)$ yields

$$\begin{aligned} I_1(t) &= (\alpha\varepsilon)^{-1} t^{d/\eta} \int_{L(S)} \mu_{m-1}(de) \\ &\qquad \int_{\mathcal{L}'_K(e) g(t)} \exp\{-y\} y^{d/(\alpha-\varepsilon)-1} \left[(1-\varepsilon)\lambda(e) g(t)\right]^{-d/(\alpha-\varepsilon)} dy, \end{aligned}$$

where

$$\mathcal{L}'_K(e) = \left\{r^{\alpha-\varepsilon}(1-\varepsilon)\lambda(e) \colon r \in \mathcal{L}_K(e)\right\}.$$

Since $\lambda(e)$ is continuous and non-vanishing on $\mathbb{S}^{m-1}$, $\lambda(e) \geq c$ for some constant $c > 0$. Then for a certain positive constant c_1

$$I_1(t) \leq c_1 g(t)^{-d/(\alpha-\varepsilon)} t^{d/\eta} \int_{L(S)} \mu_{m-1}(de) \int_{\ell'_K(e) g(t)}^{\infty} \exp\{-y\} y^{\frac{d}{\alpha-\varepsilon}-1} dy,$$

where

$$\ell'_K(e) = \inf \mathcal{L}'_K(e) = \ell_K(e)^{\alpha-\varepsilon}(1-\varepsilon)\lambda(e).$$

Lemma 4.2 yields

$$\begin{aligned} I_1(t) &\leq c_1 g(t)^{-d/(\alpha-\varepsilon)} t^{d/\eta} \mu_{m-1}(L(S)) \int_{f'_K g(t)}^{\infty} \exp\{-y\} y^{\frac{d}{\alpha-\varepsilon}-1} dy \\ &\sim c_1 g(t)^{-d/(\alpha-\varepsilon)} t^{d/\eta} \mu_{m-1}(L(S)) \exp\{-f'_K g(t)\}(f'_K g(t)) \end{aligned}$$

as $t \to \infty$, where

$$\begin{aligned} f'_K &= \inf\{\ell'_K(e) \colon e \in L(S)\} \\ &= \inf\left\{\ell_K(e)^{\alpha-\varepsilon}(1-\varepsilon)\lambda(e) \colon e \in L(S)\right\}. \end{aligned} \tag{4.9}$$

Note that

$$f'_K \geq c(1-\varepsilon)(f_K)^{\alpha-\varepsilon} > 0.$$

It follows from (4.2) that $\mu_{m-1}(L(S)) > 0$. Since $g(t)$ is regularly varying with positive index,

$$\liminf_{t\to\infty} \frac{-\log T(tK)}{g(t)} \geq \liminf_{t\to\infty} \frac{-\log I_1(t)}{g(t)} \geq f'_K.$$

Estimate the function $I_2(t)$ from (4.4) in the following way

$$I_2(t) \geq c_2 g(t)^{-d/(\alpha+\varepsilon)} t^{d/\eta} \int_{L(S)} de \int_{\mathcal{L}''_K(e) g(t)} \exp\{-y\} y^{\frac{d}{\alpha+\varepsilon}-1} dy,$$

where

$$\mathcal{L}''_K(e) = \left\{r^{\alpha+\varepsilon}(1+\varepsilon)\lambda(e) \colon r \in \mathcal{L}_K(e)\right\}.$$

The definition of $\mathcal{L}_K(e)$ yields

$$\mathcal{L}_K(e) \supset [a^{1/\eta}, b^{1/\eta}] q_S(e)^{-1/\eta}.$$

It follows from (4.8) that

$$\mathcal{L}_K(e) \supset \left[\ell_K(e), \ell_K(e) + \left(b^{1/\eta} - a^{1/\eta}\right) \tilde{q}^{-1/\eta}\right].$$

Hence for a certain $\kappa > 0$ and all e from $L(S)$ it is

$$[\ell''_K(e), \ell''_K(e) + \kappa] \subset \mathcal{L}''_K(e),$$

where

$$\ell''_K(e) = \inf \mathcal{L}''_K(e) = \ell_K(e)^{\alpha+\varepsilon}(1+\varepsilon)\lambda(e). \tag{4.10}$$

Denote

$$\begin{aligned} f''_K &= \inf\{\ell''_K(e) \colon e \in L(S)\} \\ &= \inf\left\{\ell_K(e)^{\alpha+\varepsilon}(1+\varepsilon)\lambda(e) \colon e \in L(S)\right\}. \end{aligned} \tag{4.11}$$

It follows from (**i**) and continuity of $M(e)$ that for each $\vartheta > 0$ and $e_0 \in L(S)$

$$\mu_{m-1}\left(\{e \in L(S) \colon q_S(e) \geq q_S(e_0) - \vartheta\}\right) > 0. \tag{4.12}$$

Indeed, $q_S(e_0) = \|u_0\|$ for a certain u_0 from Γ_S. By **(i)** there exists a point u_1 belonging to $M(e_1) \cap (\mathrm{Int}\Gamma_S) \cap B_{\vartheta/2}(u_0)$. It follows from continuity of the function M in the Hausdorff metric that for a certain $\delta > 0$ and each $e \in B_\delta(e_1) \subset L(S)$

$$M(e) \cap \mathrm{Int}\Gamma_S \cap B_{\vartheta/2}(u_1) \neq \emptyset.$$

Hence $q_S(e) \geq q_S(e_0) - \vartheta$ for all e from $B_\delta(e_1)$, whence (4.12) is valid.

Pick δ from the interval $(0, \kappa)$. It follows from (4.7) and (4.10) that $\ell''_K(e)$ is a continuous transformation of $q_S(e)$. From (4.12) we derive that the set

$$F_\delta = \{e \in L(S) \colon \ell''_K(e) \leq f''_K + \delta\}$$

is of positive $(m-1)$-dimensional Lebesgue measure, i.e. $\mu_{m-1}(F_\delta) > 0$.

It follows from Lemma 4.2 that for a certain constant c_2

$$\begin{aligned} I_2(t) &\geq c_2 g(t)^{-d/(\alpha+\varepsilon)} t^{d/\eta} \int_{F_\delta} \mu_{m-1}(de) \int_{(f''_K+\delta)g(t)}^{(f''_K+\kappa)g(t)} \exp\{-y\} y^{\frac{d}{\alpha+\varepsilon}-1} dy \\ &\sim c_2 g(t)^{-d/(\alpha+\varepsilon)} t^{d/\eta} \mu_{m-1}(F_\delta) \exp\left\{-(f''_K+\delta) g(t)\right\} \left(f''_K+\delta) g(t)\right)^{\frac{d}{\alpha+\varepsilon}-1} \end{aligned}$$

as $t \to \infty$. Thus

$$\limsup_{t\to\infty} \frac{-\log T(tK)}{g(t)} \leq f''_K + \delta.$$

Letting δ go to zero yields

$$f'_K \leq \liminf_{t\to\infty} \frac{-\log T(tK)}{g(t)} \leq \limsup_{t\to\infty} \frac{-\log T(tK)}{g(t)} \leq f''_K.$$

Since ε in (4.9) and (4.11) is arbitrary positive, we get

$$\lim_{t\to\infty} \frac{-\log T(tK)}{g(t)} = \Lambda(K) = \inf_{e\in L(K)} \ell_K(e)^\alpha \lambda(e). \tag{4.13}$$

It is evident that $\Lambda(K) = \Lambda(\hat{K})$ and $\Lambda(K_1) > \Lambda(K)$ if K_1 is a subset of $\mathrm{Int}K$. Thus, the conditions of Theorem 3.1 are valid for the class $\mathcal{M}$ defined in (1.2), i.e.

$$R(K) = -\log T(K) \in \mathrm{RV}(\alpha, \mathcal{M}, \Lambda, g).$$

To prove $\mathcal{K}$-convergence put $K = \{v \colon \|v\| \geq 1\}$. Then

$$\mathcal{L}_K(e) = \{r \geq 0 \colon r^\eta \|M(e)\| \geq 1\}$$

and

$$\ell_K(e) = \|M(e)\|^{-1/\eta},$$

whence

$$\Lambda(K) = \inf\left\{\|M(e)\|^{-\alpha/\eta} \lambda(e) \colon e \in \mathbb{S}^{d-1}\right\} \neq 0, \infty.$$

Theorem 3.1 and Corollary 3.2 yield $\mathcal{K}$-convergence of Y_n to the set

$$Z = Z(\Lambda; \mathcal{M}) = \left(\bigcup \{\mathrm{Int}F \colon F \in \mathcal{M}, \Lambda(F) > 1\}\right)^c.$$

For K given in (4.6) we get

$$\Lambda(K) = \inf\left\{\left(\frac{a}{q(e)}\right)^{\alpha/\eta} \lambda(e) \colon e \in L(K)\right\}.$$

If S in (4.6) tends to the single-point set $\{v\}$ for some $v \in \mathbb{S}^{d-1}$, then $\Lambda(K)$ tends to

$$\inf\left\{\left(\frac{a}{q_v(e)}\right)^{\alpha/\eta} \lambda(e) \colon e \in L(v)\right\}.$$

Hence

$$\begin{aligned} Z(\Lambda;\mathcal{M}) &= \left\{vb \colon v \in \mathbb{S}^{d-1}, b > a, \inf_{e\in L(v)} \left(\frac{a}{q_v(e)}\right)^{\alpha/\eta} \lambda(e) > 1\right\}^c \\ &= \left\{vx \colon v \in \mathbb{S}^{d-1}, 0 \le x \le a, \inf_{e\in L(v)} \left(\frac{a}{q_v(e)}\right)^{\alpha/\eta} \lambda(e) \le 1\right\} \\ &= \left\{vx \colon v \in \mathbb{S}^{d-1}, 0 \le x \le f(v)\right\}, \end{aligned}$$

where

$$\begin{aligned} f(v) &= \left[\inf\left\{\left(\frac{\lambda(e)}{q_v(e)}\right)^{\alpha/\eta} \colon e \in L(v)\right\}\right]^{-\eta/\alpha} \\ &= \sup\left\{\left(\frac{q_v(e)}{\lambda(e)}\right)^{\eta/\alpha} \colon e \in L(v)\right\}. \end{aligned}$$

If $L(v) = \emptyset$ we put $f(v) = 0$. □

NOTE. Theorem 4.1 can be easily generalized for a random vector ξ distributed in a certain cone $\mathbb{C} \subset \mathbb{R}^m$ and $M \colon \mathbb{C} \longrightarrow \mathcal{K}$.

Corollary 4.3 *Let $M(u) = \{u\}$. Then* **(i)** *is valid, $\eta = 1$, $A_1 = \{\xi\}$, $L(v) = \{v\}$ and $q_v(e) = \|e\|$. Hence $a_n^{-1}\{\xi_1, \dots, \xi_n\}$ converges in $\mathcal{K}$ to the set*

$$Z = Z(\Lambda;\mathcal{M}) = \left\{vx \colon v \in \mathbb{S}^{d-1}, x \ge 0, \lambda(vx) \le 1\right\}.$$

Note that this result coincides with the statement of Theorem 6.3 from Davis et al. (1988). Moreover, we removed the conditions of monotonicity imposed on λ in that paper.

Corollary 4.4 *Let the assumptions of Theorem 4.1 be valid and let $\lambda(e) = \lambda$ for all e from $\mathbb{S}^{d-1}$. Then the limiting set in Theorem 4.1 is equal to*

$$\left\{vx \colon v \in \mathbb{S}^{d-1}, 0 \le x \le \lambda^{-\eta/\alpha} \sup\{q_v(e) \colon e \in L(v)\}\right\}.$$

Assumption (i) is the most awkward in Theorem 4.1. Fortunately, it is valid in case the multivalued function M is defined as

$$\begin{aligned} M(x_1, \dots, x_{dn+l}) &= \operatorname{conv}\left\{(x_1, \dots, x_d), \dots, (x_{(n-1)d+1}, \dots, x_{nd})\right\} \\ &\qquad \oplus x_{dn+1} M_1 \oplus \dots \oplus x_{dn+l} M_l, \end{aligned}$$

where $M_1, \dots, M_l$ are closed subsets of $\mathbb{R}^d$. This representation covers many important examples of random sets. For example, if $n = 1$, $l = 1$, and $M_1 = B_1(0)$, then $M(x_1, \dots, x_d, x_{d+1})$ is the ball of radius x_{d+1} centered at $(x_1, \dots, x_d)$.

Note that (i) can be replaced with the condition of lower semi-continuity of the function $q_S(e)$ on $L(S)$.

Having replaced infima in (4.9) and (4.11) with essential infima, we may drop the assumption (i). Then the following theorem is valid.

Theorem 4.5 *Let a_n be defined in Theorem 4.1. Then*

$$Y_n \xrightarrow{\mathcal{F}} \left\{ vx\colon\ v \in S, S \subset \mathbb{S}^{d-1}, 0 \le x \le \operatorname*{ess\,sup}_{e \in L(S)} \left(\frac{q_S(e)}{\lambda(e)} \right)^{\eta/\alpha} \right\}$$

a.s. as $n \to \infty$.

Consider a few examples of random closed sets and almost sure limits of their normalized unions.

EXAMPLE 4.6 Let M be a non-random convex subset of $\mathbb{R}^2$ and let w_ϕ be the turn (say clockwise) to the angle ϕ. Denote

$$M(te) = t^\eta w_\phi M$$

for $t > 0$ and $e = (\cos\phi, \sin\phi)$. Then the conditions of Theorem 4.1 are valid and

$$q_v(e) = \sup\{r\colon\ rv \in w_\phi M\} = q(w_\phi^{-1} v),$$

where

$$q(u) = \sup\{r\colon\ ru \in M\}, u \in \mathbb{S}^1.$$

Similarly,

$$L(v) = \left\{e = (\cos\phi, \sin\phi) \in \mathbb{S}^1\colon\ w_\phi^{-1} v \in S_0\right\},$$

where $S_0 = \{u\|u\|^{-1}\colon u \in M \setminus \{0\}\}$. The assumption (i) is, evidently, valid. Then the limiting set in Theorem 4.1 is given by

$$Z = \left\{ xv\colon\ v \in \mathbb{S}^1, 0 \le x \le \sup_{e \in S_0} \left(\frac{q(e)}{\lambda(w_e^{-1} v)} \right)^{\eta/\alpha} \right\}.$$

If $\lambda(e) = \lambda = \text{const}$, then $Z = B_r(0)$, where

$$r = \lambda^{-\eta/\alpha} \sup\{\|x\|\colon\ x \in M\} = \lambda^{-\eta/\alpha} \|M\|.$$

EXAMPLE 4.7 Let $A_1 = B_\xi(\zeta)$ be a random ball in $\mathbb{R}^d$. Define $M(u) = B_{u_0}(u_1, \dots, u_d)$ for a vector $u = (u_0, u_1, \dots, u_d)$ from $\mathbb{R}^m$, $u_0 \ge 0$, $m = d+1$. Then $A_1 = M(\xi, \zeta)$.

The function M satisfies the conditions of Theorem 4.1 with $\eta = 1$. For any $v \in \mathbb{S}^{d-1}$ and $e = (e_0, e_1, \dots, e_d) \in \mathbb{S}^{m-1}$ we get

$$q_v(e) = \sup\{r\colon\ rv \in B_{e_0}(e_1, \dots, e_d)\}.$$

It should be noted that, in general, the evaluation of $f(v)$ in Theorem 4.1 is very complicated. If $\lambda(e) = \lambda$, then Corollary 4.4 can be applied. Since $q_v(e)$ attains its maximum for $(e_1, \dots, e_d) = tv$ and $e_0^2 + e_1^2 + \dots + e_d^2 = 1$, we get

$$\sup\{q_v(e)\colon\ e \in L(v)\} = \sup\{t + e_0\colon\ t^2 + e_0^2 = 1, t, e_0 \ge 0\} = \sqrt{2}.$$

Thus, Y_n converges to $B_r(0)$ for $r = \lambda^{-1/\alpha}\sqrt{2}$.

In more general case

$$\lambda(e) = \lambda_0(e_0) + \lambda_1(e_1, \dots, e_d).$$

(But the center and the radius of $A_1 = B_\xi(\zeta)$ are still independent.) Suppose that λ_1 is a circular symmetric function, i.e. the center ζ of A_1 has a circular symmetric distribution. Then the function $f(v)$ in Theorem 4.1 is equal to

$$f(v) = \sup\left\{\frac{e_0 + t}{(\lambda_0(e_0) + \lambda_1(tv))^{1/\alpha}}\colon\ t^2 + e_0^2 = 1\right\} = r.$$

Hence Y_n converges almost surely as $n \to \infty$ to the ball $B_r(0)$. If $\lambda_0(e_0) = \lambda_0 e_0^\alpha$ and $\lambda_1(tv) = t^\alpha \lambda_1 \|v\|^\alpha$, then

$$r = \sup\left\{\frac{e_0 + t}{(\lambda_0 e_0^\alpha + \lambda_1 t^\alpha)^{1/\alpha}}\colon\ t^2 + e_0^2 = 1\right\}.$$

EXAMPLE 4.8 Let $m = 6$, $d = 2$ and let $M(u)$ for $u = (u_1, \dots, u_6)$ be the triangle with the vertices (u_1, u_2), (u_3, u_4) and (u_5, u_6). Then the condition **(i)** is valid and the limiting set is equal to

$$Z = \left\{vx\colon\ v \in \mathbb{S}^{d-1}, 0 \le x \le \sup_{e \in L(v)} \frac{q_v(e)}{\lambda(e)^{\eta/\alpha}}\right\},$$

where

$$L(v) = \{e = (e_1, \dots, e_6) \in \mathbb{S}^{m-1}\colon\ M(e) \cap \Gamma_v \ne \emptyset\}.$$

The function $q_v(e) = q_v(e_1, \dots, e_6)$ attains its minimum for $(e_{2i-1}, e_{2i}) = t_i v$, $i = 1, 2, 3$, i.e for the degenerated triangle $M(u)$. Thus, in case $\lambda(e) = \lambda$ we get

$$\begin{aligned} f(v) &= \sup\{q_v(e)\lambda(e)^{-1/\alpha}\colon\ e \in L(v)\} \\ &= \lambda^{-1/\alpha} \sup\{\max(t_1, t_2, t_3)\colon\ t_1^2 + t_2^2 + t_3^2 = 1\} \\ &= 1. \end{aligned}$$

Hence Y_n almost surely converges to $Z = B_r(0)$, $r = \lambda^{-1/\alpha}$. For a general function λ the evaluation of $f(v)$ is much more complicated.

Chapter 6

Multivalued Regularly Varying Functions and Their Applications to Limit Theorems for Unions of Random Sets

6.1 Definition of Multivalued Regular Variation.

Multivalued functions (multifunctions) have become an important object of optimization theory and control, see Aubin and Ekeland (1984), Clarke (1983), Rockafellar and Wets (1984), Aubin and Frankowska (1990). A multivalued (or set-valued) function describe, e.g., the set of states of a control system for all admissible controls. Random multivalued functions appear in the theory of controlled random processes, random differential inclusions and stochastic optimization, see Artstein (1984), Salinetti (1987), Papageorgiou (1987), Molchanov (1991). As a rule, multivalued functions are supposed to be closed-valued.

A random closed set can be considered to be a multivalued function $A(\omega)$ defined on a probability space (Ω, σ, P), see Hiai and Umegaki (1977). Clearly, the space of elementary events Ω can be chosen to be the Euclidean space $\mathbb{R}^d$. Then the random set can be considered to be a multivalued function whose argument is a random vector. In such a way many examples of random sets can be obtained. For instance, a random ball $B_\zeta(\xi)$ in $\mathbb{R}^d$ with random center ξ and radius ζ can be represented as $M(\xi, \zeta)$, for the multivalued function $M(u, y)\colon \mathbb{R}^{d+1} \ni (u, y) \longrightarrow B_y(u)$. Some examples of homogeneous multivalued functions have been mentioned in Sections 4.4 and 5.4 in connection with limit theorems for unions of random sets.

It should be noted that random closed sets defined as multivalued functions of a random vector are easy to simulate. Then such random sets can be used as elements for the simulation of more complicated random closed sets (see Section 8.1).

In this chapter we introduce regularly varying multivalued functions and prove that the homogeneity condition in limit theorems and laws of large numbers for unions of special random sets can be replaced with the condition of the regular variation.

It will be shown also that the concept of multivalued regular variation is of use even within the frameworks of the classical theory of regularly varying functions. Namely, this concept allow to establish the inverse theorem for multivariate numerical regularly varying functions.

The main concepts of multivariate regular variation theory have been reviewed in Section 1.6. The reader is referred to Section 1.1 for definitions of the convergence in the space of closed sets.

Let $\mathbb{C}$ be a canonically closed cone in $\mathbb{R}^m$, $\mathbb{C}' = \mathbb{C} \setminus \{0\}$, and let $M: \mathbb{C} \longrightarrow \mathcal{F}$ be a multivalued function on $\mathbb{C}$ with values in the class $\mathcal{F}$ of closed subsets of $\mathbb{R}^d$. Hereafter we suppose that $M(0) = \{0\}$ and M is measurable, i.e. for any compact K the set

$$\{u \in \mathbb{C}:\ M(u) \cap K \neq \emptyset\}$$

is measurable. Note again that the dimensions d and m are not supposed to be equal.

The function M is said to be regularly varying with the limit function Φ and index α if, for any u from $\mathbb{C}'$,

$$\mathcal{F}-\lim_{t\to\infty} \frac{M(tu)}{g(t)} = \Phi(u), \tag{1.1}$$

where $\Phi(u)$ is a non-trivial closed subset of $\mathbb{R}^d$, $\Phi(u) \neq \{0\}$ for $u \neq 0$, and $g: (0,\infty) \longrightarrow (0,\infty)$ is a numerical univariate regularly varying function of index α. We then write $M \in \Pi_1(g, \mathbb{C}', \mathcal{F}, \alpha, \Phi)$ or, shortly, $M \in \Pi_1$.

If M has compact values only and (1.1) is valid for $\mathcal{K}$-limit, i.e.

$$\mathcal{K}-\lim_{t\to\infty} \frac{M(tu)}{g(t)} = \Phi(u), \tag{1.2}$$

then M is said to belong to $\Pi_1(g, \mathbb{C}', \mathcal{K}, \alpha, \Phi)$. Here $\mathcal{K}$ is the class of all compacts in $\mathbb{R}^d$.

We denote $M \in \Pi_2(g, \mathbb{C}', \mathcal{F}, \alpha, \Phi)$ if, for any sequence $u_t \in \mathbb{C}'$, such that $u_t \to u \neq 0$ as $t \to \infty$, we have

$$\mathcal{F}-\lim_{t\to\infty} \frac{M(tu)}{g(t)} = \Phi(u), \tag{1.3}$$

The class $\Pi_2(g, \mathbb{C}', \mathcal{K}, \alpha, \Phi)$ is defined similarly. As in Section 1.6, $\Pi_1 = \Pi_2$ if $m = 1$.

The classes Π_1 and Π_2 of numerical multivariate regularly varying functions were introduced in Section 1.6. We may safely think that the function $h: \mathbb{C}' \longrightarrow \mathbb{R}^1$ belongs to Π_j if and only if the one-point-valued function $M(u) = \{h(u)\} \in \Pi_j$, $j = 1, 2$.

The limiting multifunction $\Phi(u)$ is, evidently, homogeneous. Namely,

$$\Phi(su) = s^\alpha \Phi(u),$$

whatever $s > 0$ and u from $\mathbb{R}^m$ may be.

A multivalued function M is said to be $\mathcal{F}$-continuous (respectively $\mathcal{K}$-continuous) if

$$\mathcal{F}-\lim_{u\to v} M(u) = M(v)$$

(respectively for $\mathcal{K}$-limit). It was shown by Yakimiv (1981) that the limiting function λ for any numerical function from the class Π_2 is continuous. The following theorem generalizes this result for multivalued functions.

Theorem 1.1 *If M belongs to $\Pi_2(g,\mathbb{C}',\mathcal{F},\alpha,\Phi)$ (respectively $M\in\Pi_2(g,\mathbb{C}',\mathcal{K},\alpha,\Phi)$), then the multifunction Φ is $\mathcal{F}$-continuous on $\mathbb{C}'$ (respectively $\mathcal{K}$-continuous).*

PROOF. Let $u_n\to u\in\mathbb{C}'$ as $n\to\infty$. Verify the first condition of $\mathcal{F}$-convergence for the sequence $\Phi(u_n)$, $n\geq 1$. Suppose that $K\cap\Phi(u)=\emptyset$ and, moreover, $K^\varepsilon\cap\Phi(u)=\emptyset$ for some $K\in\mathcal{K}$ and $\varepsilon>0$. If $K\cap\Phi(u_n)\neq\emptyset$ for sufficiently large n, then without loss of generality we can assume that

$$\mathrm{Int}K^\varepsilon\cap\Phi(u_n)\neq\emptyset$$

for all sufficiently large n. Then

$$\mathrm{Int}K^\varepsilon\cap\frac{M(tu_n)}{g(t)}\neq\emptyset$$

for all $t\geq t_n$ and some t_n. Suppose that $t_n\uparrow\infty$ and put $u_t=u$ for $t\in[t_n,t_{n+1})$. Then (1.3) yields

$$\mathrm{Int}K^\varepsilon\cap\Phi(u)\neq\emptyset,$$

i.e. $\Phi(u)$ hits K^ε for all $\varepsilon>0$, so that $\Phi(u)$ hits K, contrary to the conjecture.

Let $\Phi(u)$ hit a certain open set G. Then $G_1\cap\Phi(u)\neq\emptyset$ for open G_1 with compact closure, such that $\bar{G}_1\subset G$. If $G\cap\Phi(u_n)=\emptyset$ for all sufficiently large n, then

$$G_1\cap\frac{M(tu_n)}{g(t)}=\emptyset$$

for $t\geq t_n$. Similar arguments as above and (1.3) yield $G_1\cap\Phi(u)=\emptyset$. Thus, $\Phi(u_n)$ $\mathcal{F}$-converges to $\Phi(u)$ as $n\to\infty$.

For M belonging to $\Pi_2(g,\mathbb{C}',\mathcal{K},\alpha,\Phi)$ the proof repeats the proof of Theorem 1 of Yakimiv (1981) reformulated for the Hausdorff distance instead of the Euclidean metric. □

Corollary 1.2 *If $M\in\Pi_2(g,\mathbb{C}',\mathcal{K},\alpha,\Phi)$, then there exists $a>0$ such that for all $b>a$ and some $C>0$*

$$M(u)\subseteq B_C(0),\ a\leq\|u\|\leq b.$$

If, additionally, $0\notin\Phi(u)$ for all $u\neq 0$, then, for a certain $\delta>0$,

$$B_\delta(0)\cap M(u)=\emptyset$$

as soon as $a\leq\|u\|\leq b$.

PROOF. Consider an arbitrary compact K missing the origin. It follows from (1.3) that there exists t_0 such that

$$M(tu)\subseteq\Phi(u)^\varepsilon g(t), t\geq t_0,$$

for $u\in\mathbb{C}'\cap\mathcal{K}$. Since g is regularly varying,

$$0<C_1=\inf_{t\in[a,b]}g(t)\leq\sup_{t\in[a,b]}g(t)=C_2<\infty$$

for some $b \geq a \geq t_0$. Hence

$$M(u) \subseteq \Phi(u\|u\|^{-\varepsilon})g(\|u\|),$$

whence the statement of Corollary 1.2 easily follows. $\square$

For *compact convex-valued* multifunctions (1.2) can be reformulated in terms of corresponding support functions.

Proposition 1.3 *Let M be a compact convex-valued multifunction. Then*

$$M \in \Pi_1(g, \mathbb{C}', \mathcal{K}, \alpha, \Phi)$$

if and only if

$$\frac{s_{M(tu)}(v)}{g(t)} \to s_{\Phi(u)}(v) \quad as \ t \to \infty$$

uniformly for v belonging to the unit sphere $\mathbb{S}^{m-1}$.

PROOF is straightforward, since the uniform convergence of support functions is equivalent to the $\mathcal{K}$-convergence of compact sets. $\square$

Consider several examples of regularly varying multivalued functions.

EXAMPLE **1.4** Let $F: \mathbb{S}^{m-1} \longrightarrow \mathcal{F}$ be a multivalued function on the unit sphere $\mathbb{S}^{m-1}$. The function M defined as

$$M(u) = \|u\|^{\alpha} F(u\|u\|^{-1}), \ u \in \mathbb{R}^m, \tag{1.4}$$

is said to be *homogeneous*. It is evident that $M \in \Pi_1(g, \mathbb{R}^m \setminus \{0\}, \mathcal{F}, \alpha, F)$ for $g(s) = s^{\alpha}$. If F is continuous in $\mathcal{F}$ (in $\mathcal{K}$), then $M \in \Pi_2$. It should be noted that the function M remains regularly varying after replacing $\|u\|^{\alpha}$ in (1.4) with $g(\|u\|)$ for any numerical regularly varying function g.

EXAMPLE **1.5** Let $m = 6$, $d = 2$, and let $M(u_1, ..., u_6)$ be the triangle in $\mathbb{R}^d$ with the vertices (u_1, u_2), (u_3, u_4) and (u_5, u_6). Then M is homogeneous and regularly varying of index 1. If $\mu(M)$ is the area of M, then the function

$$M_1(u) = \mu(M(u))^{\beta} M(u)$$

is regularly varying of index $2\beta + 1$.

EXAMPLE **1.6** Let $h_i: \mathbb{R}^m \longrightarrow \mathbb{R}^1$, $1 \leq i \leq d$, be regularly varying numerical multivariate functions from the class Π_1 on $\mathbb{C}'$, i.e.

$$\lim_{t\to\infty} \frac{h_i(tu)}{g(t)} = \phi_i(u), \ 1 \leq i \leq d, u \in \mathbb{C}'. \tag{1.5}$$

Then

$$M(u) = \{(h_1(u), \ldots, h_d(u))\} \in \Pi_1(g, \mathbb{C}', \mathcal{K}, \alpha, \Phi),$$

where $\Phi(u) = \{(\phi_1(u), \ldots, \phi_d(u))\}$ is a singleton for each u. The function M is one-point-valued function from $\mathbb{R}^m$ into the class of single-point subsets of $\mathbb{R}^d$. Note that $M \in \Pi_2$ if $h_i \in \Pi_2$, $1 \leq i \leq d$.

EXAMPLE 1.7 Let $M \in \Pi_j$, and let $h: \mathbb{R}^m \longrightarrow \mathbb{R}^1$ be a multivariate function belonging to the class Π_j of numerical functions. Then $h(u)M(u)$ is a multivalued function of the class Π_j, $j = 1, 2$.

The following lemma shows that main set-theoretic operations preserve the regular variation property.

Lemma 1.8 *Let $j = 1$ or $j = 2$, and let $M_i \in \Pi_j$, $c_i > 0$, $1 \leq i \leq p$. Then the functions*

$$\begin{aligned} M^{(1)} &= c_1 M_1 \cup \cdots \cup c_p M_p, \\ M^{(2)} &= \operatorname{conv}(M^{(1)}), \\ M^{(3)} &= c_1 M_1 \oplus \cdots \oplus c_p M_p \end{aligned}$$

belong to the same class Π_j.

PROOF follows from the continuity of the enlisted operations with respect to the convergence in $\mathcal{F}$ (in $\mathcal{K}$). □

6.2 Inversion Theorem for Multivalued Regularly Varying Functions.

The following theorem is the analog of the *inversion theorem* for numerical univariate regularly varying functions, see Section 1.6. It should be noted that this theorem cannot be generalized within the framework of numerical multivariate regularly varying functions only, since the inverse function for a *multivariate* one is necessary multivalued, see also Theorem 2.6 below.

Theorem 2.1 *Let $M \in \Pi_2(g, \mathbb{C}', \mathcal{F}, \alpha, \Phi)$ for a positive α, and let*

$$M_1(K) = \{u \in \mathbb{C}: \; M(u) \cap K \neq \emptyset\} \tag{2.1}$$

for $K \in \mathcal{K}$, $0 \notin K$. Suppose that for all u_0 from $\mathbb{C}'$ and $\varepsilon > 0$ there exists $\delta > 0$ such that

$$\Phi(u_0)^\delta \subset \bigcup_{u \in B_\varepsilon(u_0)} \frac{M(tu)}{g(t)} \tag{2.2}$$

for all sufficiently large t. Then the function M_1 is regularly varying of index $\gamma = 1/\alpha$ on the set $\mathbb{C}_a = \{u \in \mathbb{C}: \|u\| \geq a\}$ for any $a > 0$. Namely,

$$\mathcal{F} - \lim_{t \to \infty} \frac{M_1(tK_t)}{g_1(t)} \cap \mathbb{C}_a = \Phi_1(K) \cap \mathbb{C}_a. \tag{2.3}$$

Here g_1 is the asymptotically inverse function for g, see Seneta (1976) and Section 1.6,

$$\Phi_1(K) = \{u \in \mathbb{C}: \; \Phi(u) \cap K \neq \emptyset\}, \tag{2.4}$$

and $\mathcal{K}-\lim K_t = K$, $0 \notin K$. *If* $M \in \Pi_2(g, \mathbb{C}', \mathcal{K}, \alpha, \Phi)$ *and* $0 \notin \Phi(u)$, *whatever* $u \in \mathbb{C}'$ *may be, then* $\mathcal{F}$*-limit in (2.3) can be replaced with* $\mathcal{K}$*-limit. If*

$$\mathcal{K}-\lim_{t\to\infty} \frac{M(tu_t)}{g(t)} = \{0\}, \tag{2.5}$$

provided $u_t \to 0$ *as* $t \to \infty$, *then (2.3) is valid for* $a = 0$.

PROOF. It is evident that

$$\frac{M_1(tK)}{g_1(t)} = \left\{u \in \mathbb{C}\colon\ \tilde{M}_t(u) \cap f(t)K \neq \emptyset\right\}, \tag{2.6}$$

where

$$\tilde{M}_t(u) = \frac{M(g_1(t)u)}{g(g_1(t))} \tag{2.7}$$

and

$$f(t) = \frac{t}{g(g_1(t))} \to 1 \quad \text{as} \quad t \to \infty.$$

It follows from Theorem 1.1 that the set $\Phi_1(K) \cap \mathbb{C}_a$ is closed for all $a > 0$. Check the first condition of the $\mathcal{F}$-convergence in (2.3). Suppose that $K' \cap \Phi_1(K) = \emptyset$ for a certain compact $K' \subset \mathbb{C}_a$, but

$$K' \cap \frac{M_1(tK_t)}{g_1(t)} \neq \emptyset$$

for sufficiently large values of t. By (2.6) we can choose a point u_t from K' such that

$$\tilde{M}_t(u_t) \cap f(t)K_t \neq \emptyset. \tag{2.8}$$

Without loss of generality suppose $u_t \to u_0 \in K'$ as $t \to \infty$. Then (1.3) yields

$$\mathcal{F}-\lim_{t\to\infty} \tilde{M}_t(u_t) = \Phi(u_0).$$

From (2.8) we get $\Phi(u_0) \cap K^\delta \neq \emptyset$ for all $\delta > 0$, whence $\Phi(u_0) \cap K \neq \emptyset$, contrary to the conditions $u_0 \in K'$ and $K' \cap \Phi_1(K) = \emptyset$. Thus the first condition of $\mathcal{F}$-convergence is valid even without assumption (2.2).

If $a = 0$, then we have to consider additionally the case $0 \in K'$, $u_t \to u_0 = 0$ as $t \to \infty$. From (2.5) and (2.8) we get

$$0 \in (f(t)K_t)^\delta$$

for all $\delta > 0$ and sufficiently large t. However we assumed that $0 \notin K$. Thus, the condition **(F1)** from Section 1.1 is valid.

Verify the second condition of $\mathcal{F}$-convergence **(F2)** in (2.3) for $a = 0$ at once. Let $\Phi_1(K)$ have non-void intersection with a certain open set G. Since $0 \notin \Phi_1(K)$, the common point of G and $\Phi_1(K)$ is not zero, so that we can safely think that $0 \notin \bar{G}$. Let u_0 belong to $G \cap \Phi_1(K)$, and, moreover, $B_\varepsilon(u_0) \subset G$ for a certain $\varepsilon > 0$. Suppose that

$$B_\varepsilon(u_0) \cap \frac{M_1(tK_t)}{g_1(t)} = \emptyset \tag{2.9}$$

for sufficiently large t. Hence

$$\bigcup_{u\in B_\varepsilon(u_0)} \tilde{M}_t(u)\cap f(t)K_t=\emptyset. \tag{2.10}$$

From (2.2) we get

$$\Phi(u_0)^\delta \cap f(t)K_t=\emptyset.$$

Hence $\Phi(u_0)$ misses K, i.e. $u_0\notin \Phi_1(K)$. Thus, (2.3) has been proven.

Finally, suppose that $M\in \mathrm{II}_2(g,\mathbb{C}',\mathcal{K},\alpha,\Phi)$ and $0\notin\Phi(u)$ whenever $u\in\mathbb{C}'$. In order to prove $\mathcal{K}$-convergence in (2.3) we have to verify that sets $M_1(tK_t)/g_1(t)$ are contained in a certain compact for all sufficiently large t. Suppose that

$$u_t\in \frac{M_1(tK_t)}{g_1(t)}$$

for an unbounded sequence of points u_t, $t>0$. Without loss of generality suppose that the unit vector $e_t=u_t/\|u_t\|$ converges to e as $t\to\infty$. From (1.3) we get

$$\mathcal{K}-\lim_{t\to\infty}\frac{M(g_1(t)\|u_t\|e_t)}{g(g_1(t)\|u_t\|)}=\Phi(e), \tag{2.11}$$

and

$$\frac{g(g_1(t)\|u_t\|)}{g(g_1(t))}\sim \|u_t\|^\alpha \text{ as } t\to\infty.$$

Since $B_\varepsilon(0)\cap\Phi(e)=\emptyset$ for a certain $\varepsilon>0$, (2.11) yields

$$\tilde{M}_t(u_t)\subseteq \mathbb{R}^d\setminus \frac{B_\varepsilon(0)g(g_1(t)\|u_t\|)}{g(g_1(t))}$$

for sufficiently large t. Hence

$$\tilde{M}_t(u_t)\cap f(t)K_t=\emptyset,$$

contrary to the choice of u_t. □

Note that the function M_1 defined by (2.1) is said to be the *inverse* for the multifunction M, see Rockafellar and Wets (1984).

The condition (2.2) is the most awkward in Theorem 2.1. However it can be weakened a little.

Denote for any closed F and positive δ

$$[F]^\delta=\bigcup\{Fy\colon 1-\delta\le y\le 1+\delta\}. \tag{2.12}$$

Theorem 2.2 *Suppose that all conditions of Theorem 2.1 are valid except (2.2), and, for all $u_0\in\mathbb{C}'$, $\varepsilon>0$, there exists $\delta>0$ such that*

$$[\Phi(u_0)]^\delta\subseteq \bigcup_{1-\varepsilon\le q\le 1+\varepsilon}\frac{M(qu_0t)}{g(t)} \tag{2.13}$$

for all sufficiently large t. Then (2.3) is valid for $K_t=K$, $t>0$.

PROOF of Theorem 2.1 is applicable except the following implication. From (2.9) we get

$$\bigcup_{1-\varepsilon\le q\le 1+\varepsilon} \tilde{M}_t(u_0 q)\cap f(t)K = \emptyset$$

(cf. (2.10)), so that (2.13) implies

$$[\Phi(u_0)]^\delta \cap f(t)K = \emptyset.$$

Since $f(t)\to 1$ as $t\to\infty$,

$$\frac{[\Phi(u_0)]^\delta}{f(t)} \supseteq [\Phi(u_0)]^{\delta'} \supseteq \Phi(u_0)$$

for a certain $\delta'>0$ and sufficiently large t. Thus $\Phi(u_0)\cap K=\emptyset$. This fact contradicts the choice of u_0. □

Note that if $d=1$ (values of M are closed subsets of the line), then (2.2) and (2.13) are equivalent.

Reformulate (2.2) for particular functions M.

Lemma 2.3 *Let $M(u) = \{(h_1(u),\dots,h_d(u))\}$ be the single-point-valued function from Example 1.6, where h_i, $1\le i\le d$, are continuous multivariate functions from the class Π_2. If the function g in (1.5) is continuous, then (2.2) is valid.*

PROOF. Let ϕ_i, $1\le i\le d$, be limiting homogeneous functions from (1.5). Then

$$\bigcup_{u\in B_\varepsilon(u_0)} \frac{M(tu)}{g(t)} = \bigcup_{u\in B_\varepsilon(u_0)} \{(\phi_1(u),\dots,\phi_d(u)) + (\alpha_1^t(u),\dots,\alpha_d^t(u))\},$$

where

$$\alpha^t(u) = (\alpha_1^t(u),\dots,\alpha_d^t(u)) \colon \mathbb{R}^m \longrightarrow \mathbb{R}^d$$

is continuous function for any given t, and

$$\sup_{u\in B_\varepsilon(u_0)} \|\alpha^t(u)\| \to 0 \text{ as } t\to\infty.$$

Thus, (2.2) follows from

$$\bigcup_{u\in B_\varepsilon(u_0)} \{(\phi_1(u),\dots,\phi_d(u))\} \supseteq \{(\phi_1(u_0),\dots,\phi_d(u_0))\}^\delta$$

for a certain $\delta>0$. □

Lemma 2.4 *Let $M(u) = g(\|u\|)F(e_u)$, where $F\colon \mathbb{S}^{m-1}\longrightarrow \mathcal{F}$ is a multivalued function on the unit sphere and $e_u = u/\|u\|$. Then (2.13) is valid if g is a continuous numerical regularly varying function. The condition (2.5) is valid if F is bounded on S and*

$$\frac{g(tx_t)}{g(t)} \to 0 \text{ as } x_t\to 0, t\to\infty.$$

PROOF. It is evident that $\Phi(u) = \|u\|^\alpha F(e_u)$. Hence

$$\begin{aligned}\bigcup_{1-\varepsilon\leq q\leq 1+\varepsilon} \frac{M(q\|u\|t)}{g(t)} &= \bigcup_{1-\varepsilon\leq q\leq 1+\varepsilon} F(e_u)\frac{g(qt\|u\|)}{g(t)} \\ &\supseteq \bigcup_{1-\varepsilon_1\leq q\leq 1+\varepsilon_1} F(e_u)q^\alpha\|u\|^\alpha \\ &\supseteq [\Phi(u)]^\delta\end{aligned}$$

for some $\varepsilon_1 < \varepsilon$ and $\delta > 0$. Indeed,

$$\left\{\frac{g(qt\|u\|)}{g(t)}\colon\ 1-\varepsilon\leq q\leq 1+\varepsilon\right\}$$

converges in the Hausdorff metric to the set

$$\{q^\alpha\|u\|^\alpha\colon\ 1-\varepsilon\leq q\leq 1+\varepsilon\},$$

so that, for some $\varepsilon_1 > 0$ and sufficiently large t,

$$\left\{\frac{g(qt\|u\|)}{g(t)}\colon\ 1-\varepsilon\leq q\leq 1+\varepsilon\right\}\supseteq\{q^\alpha\|u\|^\alpha\colon\ 1-\varepsilon_1\leq q\leq 1+\varepsilon_1\}. \quad \square$$

Note also that all functions from Lemma 1.8 satisfy the conditions (2.2) or (2.13) in case all their components M_i, $1\leq i\leq p$, satisfy the same condition.

Without (2.2) or (2.13) the following result is valid.

Corollary 2.5 *If the conditions of Theorem 2.1 are valid except (2.2), then (2.3) is replaced with*

$$\mathcal{F}\text{-}\limsup_{t\to\infty}\frac{M_1(tK_t)}{g_1(t)}\cap \mathbb{C}_a \ \subseteq\ \Phi_1(K)\cap\mathbb{C}_a. \tag{2.14}$$

PROOF follows from Lemma 1.1.1 (for definition of the upper limit in $\mathcal{F}$ see Section 1.1). $\square$

Note that the closed sets K, K_t, $t > 0$ in Theorem 2.1 are allowed to be non-compact, provided $\rho_H(K, K_t) \to 0$ as $t\to\infty$.

Now consider a particular case of Theorem2.1, which, in fact, is the inversion theorem for multivariate regularly varying functions.

Theorem 2.6 *Let $h\colon \mathbb{C}\longrightarrow\mathbb{R}^1$ be a continuous regularly varying numerical function ($h\in\Pi_2$) with the limiting function ϕ and index $\alpha > 0$. Suppose that the corresponding norming function g in (1.5) is continuous. Define for any $x\geq 0$, $a > 0$*

$$M_1(x) = \{u\in\mathbb{C}\colon\ \|u\|\geq a, h(u)\geq x\}. \tag{2.15}$$

Then

$$M_1\in\Pi_2(g_1,(0,\infty),\mathcal{F},\gamma,\Phi_1),$$

where g_1 is the asymptotically inverse function for g, $\gamma = 1/\alpha$ and

$$\Phi_1(x) = \{u\in\mathbb{C}\colon\ \phi(u)\geq x\}.$$

PROOF. Since M_1 is defined on $(0,\infty)$, the corresponding classes Π_1 and Π_2 coincide. Let us apply Theorem 2.2 to the one-point-valued function $M(u) = \{h(u)\}$, $u \in \mathbb{C}$. Then (2.13) is equivalent to

$$[1-\delta, 1+\delta] \subseteq \left\{ \frac{h(qut)}{g(t)\phi(u)} \colon 1-\varepsilon \le q \le 1+\varepsilon \right\}. \tag{2.16}$$

The inclusion (2.16), in turn, follows from continuity of h, ϕ and g. Note that the continuity condition can be replaced with a certain analog of coordinate-wise monotonicity.

The function h can be redefined on $\mathbb{C} \setminus \mathbb{C}_a$ to ensure (2.5). Indeed, (2.5) is valid in case $u_t t$ is divided from the origin. Otherwise, letting h be equal to zero on $\mathbb{C} \setminus \mathbb{C}_a$ ensures (2.5). It follows from Theorem 2.1 that

$$\mathcal{F} - \lim_{t\to\infty} \frac{M_1(xt)}{g_1(t)} = \Phi_1(x),$$

i.e. M_1 is regularly varying. $\square$

Note that we can construct the next inverse function to M defined by (2.15). This function M_2 is defined as

$$M_2(K) = [0, \sup_{u\in K} h(u)], \ \ K \subset \mathbb{C}.$$

Naturally, the function M_2 is regularly varying too.

6.3 Integrals on Multivalued Regularly Varying Functions.

Results, concerning asymptotic properties of integrals of regularly varying functions, constitute a large part of classical regular variation theory. In this section we consider asymptotic properties of an integral, whose *domain of integration* is a certain multivalued regularly varying function.

Theorem 3.1 *Let $L\colon \mathbb{G}' \to \mathbb{R}^1$ be a slowly varying function (i.e. $L \in \mathfrak{W}_2$, see Section 1.6), and let $\phi\colon \mathbb{G}' \to \mathbb{R}^1$ be a continuous homogeneous function of index $\alpha - d$, $\alpha < 0$, where $\mathbb{G}' = \mathbb{G} \setminus \{0\}$, $\mathbb{G}$ is a cone in $\mathbb{R}^d$. Furthermore, let*

$$M\colon \mathbb{R}^1 \longrightarrow \mathcal{F}(\mathbb{G}) = \{F \in \mathcal{F}\colon F \subseteq \mathbb{G}'\}$$

be a multivalued function, whose values are closed subsets of $\mathbb{G}'$. Suppose that for some canonically closed set D, missing the origin, and, for every compact K,

$$\inf\left\{\varepsilon > 0\colon D^{-\varepsilon} \cap K \subset \frac{M(t)}{g(t)} \cap K \subset D^{\varepsilon} \cap K\right\} \to 0 \ \ as \ t \to \infty, \tag{3.1}$$

where $g\colon (0,\infty) \longrightarrow (0,\infty)$ is a regularly varying function of index $\gamma > 0$. Then, for any $e \in \mathbb{G}'$,

$$\int_{M(t)} \phi(u)L(u)du \sim L(g(s)e)g(s)^{\alpha} \int_D \phi(u)du \ \ as \ s \to \infty. \tag{3.2}$$

First, derive a lemma concerning integrals of regularly varying functions.

Lemma 3.2 *Let F be a closed subset of $\mathbb{R}^d \setminus \{0\}$, and let ϕ, L be defined in Theorem 3.1. Then, for any e from $\mathbb{R}^d \setminus \{0\}$,*

$$\int_F \phi(u)L(xu)du \sim L(xe)\int_F \phi(u)du \quad as \ x \to \infty. \tag{3.3}$$

PROOF. Evidently,

$$c = \inf\{\|x\|: \ x \in F\} > 0.$$

It was proven in de Haan and Resnick (1987) that, for any $\varepsilon > 0$, there exists x_0 such that for $x \geq x_0$ and $\|x\| \geq c$ it is

$$(1-\varepsilon)\|u\|^{-\varepsilon} \leq \frac{L(xu)}{L(xe)} \leq (1+\varepsilon)\|u\|^{\varepsilon}.$$

Let $\Delta \in (0,1)$ be specified. Since $\alpha < 0$, we can choose $R > 0$ such that

$$\begin{aligned}
\int_{F\cap B_R^c(0)} \phi(u)\|u\|^{\varepsilon}du &< \Delta\int_F \phi(u)du,\\
\int_{F\cap B_R(0)} \phi(u)du &> (1-\Delta)\int_F \phi(u)du.
\end{aligned}$$

Thus

$$\begin{aligned}
\int_{F\cap B_R(0)} \phi(u)(1-\varepsilon)\|u\|^{-\varepsilon}L(xe)du &\leq \int_F \phi(u)L(xu)du\\
&\leq L(xe)(1+\varepsilon)\int_{F\cap B_R(0)} \phi(u)\|u\|^{\varepsilon}du\\
&\qquad +\Delta\int_F \phi(u)du.
\end{aligned}$$

Hence

$$(1-\Delta)(1-\varepsilon)R^{-\varepsilon} \leq \frac{\int_F \phi(u)L(xu)du}{L(xe)\int_F \phi(u)du} \leq (1+\varepsilon)R^{\varepsilon} + \Delta.$$

Since ε may be chosen sufficiently small, (3.3) is valid. □

PROOF OF THEOREM 3.1. If $x = g(t)$, then

$$\int_{M(t)} \phi(u)L(u)du = x^{\alpha}\int_{M(t)/x} \phi(u)L(ux)du.$$

Since $0 \notin D$, it is

$$\int_{M(t)/x} \phi(u)\|u\|^{\eta}du < \infty$$

for sufficiently large t and any $\eta \in (0, -\alpha)$. Hence, for all $R > 0$,

$$\begin{aligned}
x^{\alpha}\int_{D^{-\epsilon}\cap B_R(0)} \phi(u)L(ux)du &\leq x^{\alpha}\int_{\frac{M(t)}{x}\cap B_R(0)} \phi(u)L(ux)du\\
&\leq x^{\alpha}\int_{D^{\epsilon}} \phi(u)L(ux)du
\end{aligned} \tag{3.4}$$

Denote

$$\Lambda(F) = \int_F \phi(u)du, \; F \in \mathcal{F}.$$

Lemma 3.2 and (3.4) yield

$$\frac{\Lambda(D^{-\varepsilon} \cap B_R(0))}{\Lambda(D)} \leq \lim_{t\to\infty} \frac{\int_{M(t)} \phi(u)L(u)du}{L(xe)x^\alpha \Lambda(D)} \leq \frac{\Lambda(D^\varepsilon)}{\Lambda(D)}.$$

To obtain (3.2) we have to put $R \to \infty$, $\varepsilon \downarrow 0$ and use continuity of ϕ. □

Note. The condition (3.1) is more restrictive than

$$\mathcal{F} - \lim_{t\to\infty} \frac{M(t)}{g(t)} = D.$$

Nevertheless, for *convex-valued* multifunctions these conditions are equivalent.

Corollary 3.3 *Let $M \in \Pi_2(g, \mathbb{C}', \mathcal{K}, \alpha, \Phi)$ be a convex-valued multifunction, and let the functions ϕ and L satisfy the conditions of Theorem 3.1. Then*

$$H(v) = \int_{M(v)} \phi(u)L(u)du$$

is a regularly varying multivariate function from the class Π_2. In particular $\mu(M(v)) \in \Pi_2$, where μ is the Lebesgue measure.

6.4 Limit Theorems for Unions: Multivalued Functions Approach.

In this section we apply the above mentioned results to limit theorems for unions of random sets considered in Sections 4.1 and 4.4.

Let ξ be a random point in $\mathbb{C} \subseteq \mathbb{R}^m$ having the density f. Suppose that $f \in \Pi_2$ on $\mathbb{C}' = \mathbb{C} \setminus \{0\}$ and ind$f = \alpha - m$ for a certain negative α. Then $f = \phi L$, where ϕ is a homogeneous continuous function of the same index, and L is a slowly varying function on $\mathbb{C}'$.

Furthermore, let M be a multivalued function from the class $\Pi_2(g, \mathbb{C}', \mathcal{K}, \eta, \Phi)$, $\eta > 0$. Then $A = M(\xi)$ is a random compact set. Consider its independent copies $A_1, A_2, \ldots$ and define

$$a_n^{-1} X_n = a_n^{-1}(A_1 \cup \cdots \cup A_n),$$

for the norming constants a_n, $n \geq 1$, given by

$$a_n = \sup\left\{g(s)\colon\ s^\alpha L(se) \geq 1/n\right\}, \tag{4.1}$$

for a certain $e \in \mathbb{C}'$.

In this section we investigate the weak convergence of random closed sets $a_n^{-1}X_n$, $n \geq 1$. As it was stated in Section 1.4, the weak convergence of random sets is equivalent to the pointwise convergence of the corresponding capacity functionals

$$T_n(K) = \mathbf{P}\left\{a_n^{-1}X_n \cap K \neq \emptyset\right\}.$$

Moreover, the pointwise convergence of T_n on the class $\mathcal{K}_{ub}$ of all finite unions of balls ensures the weak convergence of the random sets in question.

Hereafter in this section we suppose that the above mentioned conditions on the function M, the random variable ξ and its density are satisfied.

Theorem 4.1 *Suppose that, for every $u_0 \in \mathbb{C}'$, positive r and K from $\mathcal{K}_{ub}$, the conditions*

$$\Phi(u_0) \cap K \neq \emptyset \text{ and } \Phi(u_0) \cap \mathrm{Int}K = \emptyset,$$

yield the existence of some points u_1 and u_2 belonging to $B_r(u_0)$ such that

$$\Phi(u_1) \cap K = \emptyset \text{ and } \Phi(u_2) \cap \mathrm{Int}K \neq \emptyset.$$

In addition, let (2.5) be valid for the multivalued function M. Then $a_n^{-1}X_n$ converges weakly to the random closed set X with the capacity functional $\tilde{T}$ given by

$$\tilde{T}(K) = \begin{cases} 1 - \exp\left\{-\int_{\Phi_1(K)} \phi(u)du\right\} & , \quad 0 \notin K, \\ 1 & , \quad \text{otherwise} \end{cases},$$

where Φ_1 is the inverse function to Φ, see (2.4).

PROOF. Since $a_n \to \infty$ as $n \to \infty$, the origin is a fixed point of the limiting random set, i.e. $\tilde{T}(K) = 1$ as soon as $0 \in K$.

Let us verify the pointwise convergence of capacity functionals. Due to Theorem 4.1.1, it suffices to show that the function

$$\begin{aligned} \tau_K(x) &= \mathbf{P}\{A \cap xK \neq \emptyset\} \\ &= \mathbf{P}\{M(\xi) \cap xK \neq \emptyset\} \end{aligned}$$

is regularly varying at infinity for any K from $\mathcal{K}_{ub}$. Using the notations of Theorem 2.1 we get

$$\begin{aligned} \tau_K(x) &= \mathbf{P}\{\xi \in M_1(xK)\} \\ &= \int_{M_1(xK)} \phi(u)L(u)du. \end{aligned}$$

It follows from Corollary 2.5 that

$$\inf\left\{\varepsilon > 0 \colon \frac{M_1(xK)}{g_1(x)} \cap B_R(0) \subseteq \Phi_1(K)^{\varepsilon}\right\} \to 0 \text{ as } x \to \infty \tag{4.2}$$

for any $R > 0$.

Let us show that

$$\inf\left\{\varepsilon > 0 \colon \frac{M_1(xK)}{g_1(x)} \supseteq \Phi_1(K)^{-\varepsilon} \cap B_R(0)\right\} \to 0 \text{ as } x \to \infty. \tag{4.3}$$

Suppose that

$$\Phi_1(K)^{-\varepsilon} \cap B_R(0) \not\subset \frac{M_1(xK)}{g_1(x)}$$

for some $\varepsilon > 0$ and $x = x_k$, $k \geq 1$, where $x_k \to \infty$ as $k \to \infty$. Then pick points u_k, $k \geq 1$, such that

$$u_k \in \left(\Phi_1(K)^{-\varepsilon} \cap B_R(0)\right) \setminus \frac{M_1(x_k K)}{g_1(x_k)}.$$

Let $u_k \to u_0 \in \mathrm{Int} B_R(0)$ as $k \to \infty$. Since $0 \notin \Phi_1(K)$, we get $u_0 \neq 0$. Thus, $\Phi(u) \cap K \neq \emptyset$ whenever $u \in B_{\varepsilon/2}(u_0)$. On the other hand,

$$\tilde{M}_{x_k}(u_k) \cap K = \emptyset,$$

i.e. $\Phi(u_0) \cap \mathrm{Int}K = \emptyset$. On the contrary, by the assumption, $\Phi(u_1)$ misses K for a certain u_1 belonging to $B_{\varepsilon/2}(u_0)$. Thus, (4.3) is valid.

Let us show that $\Phi_1(K)$ is canonically closed. If $\Phi(u_0) \cap \mathrm{Int}K \neq \emptyset$, then $u_0 \in \mathrm{Int}\Phi_1(K)$ by Theorem 1.1. Let

$$\Phi(u_0) \cap \mathrm{Int}K = \emptyset \text{ and } \Phi(u_0) \cap K \neq \emptyset.$$

In view of the assumption of Theorem, there exists a sequence of points u_k, $k \geq 1$, such that $u_k \to u_0$ as $k \to \infty$ and

$$\Phi(u_k) \cap \mathrm{Int}K \neq \emptyset, \; k \geq 1.$$

Hence any point u_0 from $\Phi_1(K)$ is a limit of a sequence of points from $\mathrm{Int}\Phi_1(K)$, i.e. the set $\Phi_1(K)$ is canonically closed.

It follows from (4.2), (4.3) and Theorem 3.1 that

$$\tau_K(x) \sim \nu(x) = L(g_1(x)e)(g_1(x))^{\alpha} \int_{\Phi_1(K)} \phi(u)du \text{ as } x \to \infty.$$

The function $\nu(x)$ is regularly varying of index α/η. It follows from Theorem 4.1.1 that

$$\begin{aligned} \lim_{n\to\infty} T_n(K) &= \tilde{T}(K) \\ &= 1 - \exp\left\{ - \lim_{n\to\infty} \left(\frac{a_n(K)}{a_n} \right)^{-\alpha/\eta} \right\}, \end{aligned}$$

where

$$a_n(K) = \sup\{s \colon \tau_K(s) \geq 1/n\}.$$

Let us define

$$\begin{aligned} \Lambda(K) &= \int_{\Phi_1(K)} \phi(u)du, \\ y(x) &= \left(g_1(x)^{\alpha} L(g_1(x)e)\right)^{-1}. \end{aligned}$$

Then

$$a_n = \sup\{x \colon y(x) \leq n\},$$

and, for sufficiently large n,

$$\begin{aligned} a_n(K) &\leq \sup\left\{x\colon\ g_1(x)^\alpha L(g_1(x)e)(1+\beta)\Lambda(K) \geq \frac{1}{n}\right\} \\ &= \sup\left\{x\colon\ y(x) \leq n\Lambda(K)(1+\beta)\right\}. \end{aligned}$$

Let $\bar{y}$ be the asymptotically inverse function for y. Since y is regularly varying of index $(-\alpha/\eta)$, we get

$$\begin{aligned} \lim_{n\to\infty}\left(\frac{a_n(K)}{a_n}\right)^{-\alpha/\eta} &\leq \lim_{n\to\infty}\left(\frac{\bar{y}(n\Lambda(K)(1+\beta))}{\bar{y}(n)}\right)^{-\alpha/\eta} \\ &= \Lambda(K)(1+\beta), \end{aligned}$$

for any $\beta > 0$. The estimates from below are obtained similarly. Thus, the formula for $\tilde{T}$ in Theorem 4.1 is valid. □

NOTE. We can choose instead of $\mathcal{K}_{ub}$ another class $\mathcal{M}$ determining the weak convergence, see Section 1.4 and Norberg (1984). The statement of Theorem 4.1 is also true even in case the conditions is valid for the class $\mathcal{M}'$ such that

$$K^{-\varepsilon} \subset K_1 \subset K \subset K_2 \subset K^\varepsilon$$

for any K from $\mathcal{M}$, $\varepsilon > 0$ and some K_1, K_2 from $\mathcal{M}'$.

Note also that the conditions of Theorem 4.1 are valid for all examples of regularly varying multifunctions from Section 6.1.

EXAMPLE **4.2** Let $M = g(\|u\|)B_r(e_u)$, where $r > 0$, $e_u = u/\|u\|$ and g is a regularly varying univariate function of index $\eta > 0$ such that

$$\frac{g(x_t t)}{g(t)} \to 0 \text{ as } x_t \to 0, t \to \infty.$$

Furthermore, let ξ be a random vector which satisfies the conditions of Theorem 4.1. Then the RACS $a_n^{-1}X_n$ converges weakly to the random closed set X with the capacity functional

$$\tilde{T}(K) = 1 - \exp\left\{-\int_{\mathbb{S}^{m-1}} \phi(e)de \int_{F_K(e)} x^{\alpha-1}dx\right\},\ 0 \notin K,$$

where

$$F_K(e) = \{x > 0\colon\ x^\eta B_r(e) \cap K \neq \emptyset\}.$$

Suppose that the distribution of ξ is spherically symmetric, i.e. $\phi(e) = C$ for all e belonging to $\mathbb{S}^{m-1}$. Then

$$\tilde{T}(K) = 1 - \exp\left\{-C\int_0^\infty x^{\alpha-1}\mu_{m-1}\left(\mathbb{S}^{m-1} \cap (K/x^\eta)\right)dx\right\},\ 0 \notin K,$$

where μ_{m-1} is the Lebesgue measure on $\mathbb{S}^{m-1}$.

Chapter 7

Probability Metrics in the Space of Random Sets Distributions

7.1 Definitions of Probability Metrics.

In this chapter we discuss probability metrics in the space of random closed sets distributions. *Probability metrics method* and its applications to limit theorems were elaborated by Zolotarev (1986), Kalashnikov and Rachev (1988), Rachev (1991). This method is developed mostly for distributions of random variables. There are many examples of probability metrics for random variables and inequalities between these metrics.

The probability metrics method enables to prove limit theorems for the most convenient metric. Afterwards, estimates of the speed of convergence are reformulated for other metrics by the instrumentality of inequalities between metrics. Sometimes this method allows to drop the condition of the uniform smallness of summands in limit theorems, i.e. to prove "non-classical" versions of limit theorems.

The *probability metric* $\mathfrak{m}(\xi,\eta)$ is a numerical function on the space of distributions of random elements. It satisfies the following conditions:

1. $\mathfrak{m}(\xi,\eta)=0$ implies $\mathbf{P}\{\xi=\eta\}=1$.
2. $\mathfrak{m}(\xi,\eta)=\mathfrak{m}(\eta,\xi)$.
3. $\mathfrak{m}(\xi,\eta)\leq\mathfrak{m}(\xi,\zeta)+\mathfrak{m}(\zeta,\eta)$.

In this section several probability metrics for random sets are defined. They enable to determine distances between random sets distributions. Later on their applications to limit theorems for unions are considered.

Since a random set is an $\mathcal{F}$-valued random element, probability metrics for random sets can be defined by specializing general metrics for the case of random elements in the space $\mathcal{F}$ furnished with σ-algebra σ and the Hausdorff distance ρ_H.

In such a way the Levy-Prohorov metric can be defined, because its form does not depend essentially on the structure of the setting space. We can also define the metric K_H as

$$K_H(X,Y)=\inf\{\varepsilon>0\colon\ \mathbf{P}\{\rho_{\mathrm{H}}(X,Y)>\varepsilon\}<\varepsilon\},$$

where X and Y are random compact sets. It can be shown that K metrizes the convergence of random compact sets in probability with respect to the Hausdorff metric. The analog of so-called "engineering" metric (see Zolotarev, 1986) is defined as

$$I_H(X,Y) = \mathbf{E}\rho_{\mathrm{H}}(X,Y).$$

The enlisted metrics are *composite*, i.e. their values depend on the mutual distributions of X and Y. It is well-known that *simple* metrics are more convenient, since they can be naturally applied to limit theorems. A probability metric is said to be simple if its values depend only on marginal distributions of random elements (random sets).

Many interesting simple metrics for random variables are defined by the corresponding densities or characteristic functions. Unfortunately, they cannot be reformulated for random sets directly, since the space $\mathcal{F}$ of closed sets does not admit a group operation and there are not analogues of the Lebesgue measure and densities for $\mathcal{F}$-valued random elements (random sets).

Another approach is based on the notion of selector for random sets, see Wagner (1979). The random element ξ is said to be a *selector* of X if $\xi \in X$ almost surely. We then write $\xi \in \mathcal{S}(X)$. If the random closed set X is nonempty almost surely, then the class $\mathcal{S}(X)$ is non-void too. Moreover, X coincides with the closure of a certain countable collection of its selectors. This collection is called the *Castaign representation* of X.

Let $\mathfrak{m}$ be a probability metric on the space of distributions of random vectors in $\mathbb{R}^d$. Then the metric $\mathfrak{m}_H$ on the space of random sets distributions is introduced in the same way as the Hausdorff metric ρ_{H} is defined by the Euclidean metric ρ in $\mathbb{R}^d$. Put

$$\mathfrak{m}_H(X,Y) = \max\left\{ \sup_{\xi\in\mathcal{S}(X)} \inf_{\eta\in\mathcal{S}(Y)} \mathfrak{m}(\xi,\eta),\ \sup_{\eta\in\mathcal{S}(Y)} \inf_{\xi\in\mathcal{S}(X)} \mathfrak{m}(\xi,\eta) \right\}.$$

It is easy to show that $\mathfrak{m}_H$ is a probability metric on the space of random sets distributions. Moreover, $\mathfrak{m}_H$ inherits the homogeneous property of $\mathfrak{m}$. Namely, if $\mathfrak{m}$ is homogeneous of degree γ, i.e.

$$\mathfrak{m}(c\xi, c\eta) = |c|^\gamma \mathfrak{m}(\xi,\eta), c \neq 0,$$

then $\mathfrak{m}_H$ is homogeneous too. Indeed, the class $\mathcal{S}(cX)$ coincides with $c\mathcal{S}(X)$, whatever $c \neq 0$ may be.

EXAMPLE 1.1 Let $\mathfrak{m}$ be the simple engineering metric, i.e. $\mathfrak{m}(\xi,\eta) = \rho(\mathbf{E}\xi, \mathbf{E}\eta)$. Then

$$\begin{aligned}
\mathfrak{m}_H(X,Y) &= \max\left\{ \sup_{\xi\in\mathcal{S}(X)} \inf_{\eta\in\mathcal{S}(Y)} \rho(\mathbf{E}\xi,\mathbf{E}\eta),\ \sup_{\eta\in\mathcal{S}(Y)} \inf_{\xi\in\mathcal{S}(X)} \rho(\mathbf{E}\xi,\mathbf{E}\eta) \right\} \\
&= \max\left\{ \sup_{x\in\mathbf{E}X} \inf_{y\in\mathbf{E}Y} \rho(x,y),\ \sup_{y\in\mathbf{E}Y} \inf_{x\in\mathbf{E}X} \rho(x,y) \right\} \\
&= \rho_{\mathrm{H}}(\mathbf{E}X, \mathbf{E}Y),
\end{aligned}$$

i.e. in this case $\mathfrak{m}_H$ coincides with the Hausdorff distance between the corresponding expectations of random sets. As in Section 2.1, $\mathbf{E}X$ designates the Aumann expectation of the random set X.

EXAMPLE 1.2 Let $\mathfrak{m}(\xi,\eta) = \mathbf{E}\rho(\xi,\eta)$. Then

$$\mathfrak{m}_H(\xi,\eta) = \mathbf{E}\rho_{\mathrm{H}}(X,Y).$$

Unfortunately, for a more complicated metric $\mathfrak{m}$ the evaluation of $\mathfrak{m}_H$ for random sets is very difficult, since the class of all selectors is very large even for simple random sets.

EXAMPLE 1.3 Let $X = \{\xi_1,\xi_2\}$ and $Y = \{\eta_1,\eta_2\}$ be two-point random sets in $\mathbb{R}^1$. Clearly, the class $\mathcal{S}(X)$ consists of trivial selectors ξ_1 and ξ_2, as well as all selectors defined as $\xi_{f(\xi_1,\xi_2)}$, where $f: \mathbb{R}^2 \longrightarrow \{1,2\}$ is a Borel function. Then

$$\begin{aligned} \mathbf{P}\left\{\xi_{f(\xi_1,\xi_2)} < x\right\} &= \mathbf{P}\left\{\xi_1 < x, (\xi_1,\xi_2) \in F\right\} + \mathbf{P}\left\{\xi_2 < x, (\xi_1,\xi_2) \notin F\right\} \\ &= \mathbf{P}\left\{(\xi_1,\xi_2) \in F_x\right\}, \end{aligned}$$

where

$$\begin{aligned} F &= \left\{(x,y) \in \mathbb{R}^2 \colon f(x,y) = 1\right\}, \\ F_x &= (F \cap ((-\infty,x) \times \mathbb{R})) \cup (F^c \cap (\mathbb{R} \times (-\infty,x))). \end{aligned}$$

Let $\mathfrak{m}$ be the uniform distance between random variables. It is defined as the uniform distance between the corresponding distributions functions, see Zolotarev (1986). Then the distance $\mathfrak{m}_H$ between X and Y is evaluated by

$$\begin{aligned} &\mathfrak{m}_H(X,Y) = \\ &= \max\left\{ \sup_{F\in\mathcal{B}} \inf_{G\in\mathcal{B}} \sup_{-\infty<x<\infty} \left|\mathbf{P}\left\{(\xi_1,\xi_2) \in F_x\right\} - \mathbf{P}\left\{(\eta_1,\eta_2) \in G_x\right\}\right|; \right. \\ &\qquad \left. \sup_{G\in\mathcal{B}} \inf_{F\in\mathcal{B}} \sup_{-\infty<x<\infty} \left|\mathbf{P}\left\{(\xi_1,\xi_2) \in F_x\right\} - \mathbf{P}\left\{(\eta_1,\eta_2) \in G_x\right\}\right| \right\}, \end{aligned}$$

where $\mathcal{B}$ designates the class of Borel subsets of $\mathbb{R}^2$. Thus, the evaluation of $\mathfrak{m}_H$ even for simple $\mathfrak{m}$ and two-point random sets is very complicated.

Meaningful generalizations of famous probability metrics can be obtained by replacing distribution functions in their definitions with capacity functionals of random sets. The capacity functional of X is defined as $T_X(K) = \mathbf{P}\{X \cap K \neq \emptyset\}$ for K belonging to the class $\mathcal{K}$ of all compact subsets of $\mathbb{R}^d$. Sometimes we consider the restriction of T_X on a certain sub-class $\mathcal{M} \subseteq \mathcal{K}$.

The *uniform distance* between the random sets X and Y is defined as

$$\mathfrak{r}(X,Y;\mathcal{M}) = \sup\left\{|T_X(K) - T_Y(K)| \colon K \in \mathcal{M}\right\}, \tag{1.1}$$

where $\mathcal{M}$ is a subclass of $\mathcal{K}$. The *Levy metric* is defined as follows (see also Rachev, 1986, and Baddeley, 1991)

$$\mathfrak{L}(X,Y;\mathcal{M}) = \inf\left\{\varepsilon > 0 \colon T_X(K) \leq T_Y(K^\varepsilon) + \varepsilon, T_Y(K) \leq T_X(K^\varepsilon) + \varepsilon\right\}, \tag{1.2}$$

where K^ε is the ε-envelope of K, see Section 1.1.

Hereafter we omit $\mathcal{M}$ if $\mathcal{M} = \mathcal{K}$, i.e. $\mathfrak{r}(X,Y) = \mathfrak{r}(X,Y;\mathcal{K})$ and $\mathfrak{L}(X,Y) = \mathfrak{L}(X,Y;\mathcal{M})$ etc.

We prove below that the Levy metric determines the *weak convergence* of random sets, see Section 1.4 for the notion of the weak convergence. The RACS X_n with the capacity functional T_n converges to the RACS X, having the capacity functional T, if

$$T_n(K) \to T(K) \text{ as } n \to \infty \tag{1.3}$$

for all K belonging to the class

$$\mathcal{S}_T = \{K \in \mathcal{K}: T(K) = T(\mathrm{Int}K)\}.$$

The class $\mathcal{M} \subseteq \mathcal{K}$ is said to *determine the weak convergence* of random sets if the pointwise convergence (1.3) for all K belonging to $\mathcal{M} \cap \mathcal{S}_T$ yields the weak convergence of random sets. It was noted in Section 1.4 that such classes as the class $\mathcal{K}_{ub}$ of finite unions of balls or the class $\mathcal{K}_{up}$ of finite unions of parallelepipeds determine the weak convergence of random closed sets.

For each compact K_0 put

$$\mathcal{M}(K_0) = \{K \in \mathcal{M}: K \subseteq K_0\}.$$

Theorem 1.4 *Let the class $\mathcal{M} \subseteq \mathcal{K}$ determine the weak convergence of random sets, and let* $\mathrm{Int}K$ *for each $K \in \mathcal{M}$ be equal to the limit of an increasing sequence $\{K_n, n \geq 1\} \subset \mathcal{M}$. Then a sequence X_n, $n \geq 1$, of random sets converges weakly to X if and only if, for each $K_0 \in \mathcal{K}$,*

$$\mathfrak{L}(X_n, X; \mathcal{M}(K_0)) \to 0 \quad \textit{as } n \to \infty.$$

PROOF. *Sufficiency.* Let $\mathfrak{L}(X_n, X; \mathcal{M}(K_0)) \to 0$ as $n \to \infty$, and let $K \in \mathcal{M}(K_0) \cap \mathcal{S}_T$. It follows from (1.2) that

$$T(K) \leq T_n(K^{\varepsilon_n}) + \varepsilon_n \text{ and } T_n(K) \leq T(K^{\varepsilon_n}) + \varepsilon_n, \quad n \geq 1, \tag{1.4}$$

where $\varepsilon_n \downarrow 0$ as $n \to \infty$. It follows from the conditions on $\mathcal{M}$ imposed in Theorem that

$$T(K_n) \uparrow T(\mathrm{Int}K) = T(K) \tag{1.5}$$

for a sequence K_n, $n \geq 1$, from $\mathcal{M}$. Having renumbered the sequence K_n, $n \geq 1$, one can ensure that

$$K_n^{\varepsilon_n} \subseteq K, n \geq 1.$$

Since (1.4) is valid on $\mathcal{M}(K_0)$,

$$T_n(K_n) \leq T_n(K_n^{\varepsilon_n}) + \varepsilon_n \leq T_n(K) + \varepsilon_n.$$

Thus

$$T(K) - \varepsilon_n - (T(K) - T(K_n) \leq T_n(K) \leq T(K) + \varepsilon_n + (T(K^{\varepsilon_n}) - T(K)).$$

Upper semi-continuity of T and (1.5) yield $T_n(K) \to T(K)$ as $n \to \infty$.

Necessity. Let X_n converges weakly to X. Then (1.3) is valid, whatever K belonging to $\mathcal{M} \cap \mathcal{S}_T$ may be.

Let $\varepsilon > 0$ and $K_0 \in \mathcal{K}$ be specified. Consider compacts $K_1, \ldots, K_m$, which form the ε-net of $\mathcal{M}(K_0)$ in the Hausdorff metric ρ_{H}. It is easy to show that $K_i^{r_i}$ belongs to $\mathcal{S}_T$ for a certain $r_i \in [\varepsilon, 2\varepsilon]$, $1 \le i \le m$. It follows from (1.3) that for a certain integer n_0 and every $n \ge n_0$, $1 \le i \le m$,

$$|T_n(K_i^{r_i}) - T(K_i^{r_i})| \le \varepsilon.$$

Let $K \in \mathcal{M}(K_0)$, and let K_j be its nearest neighbor from the chosen ε-net. Then, for all $n \ge n_0$,

$$\begin{aligned} T_n(K) \le T_n(K_j^{\varepsilon}) &\le T_n(K_j^{r_j}) \\ &\le T(K_j^{r_j}) + \varepsilon \le T(K^{3\varepsilon}) + 3\varepsilon. \end{aligned} \tag{1.6}$$

Similarly,

$$T(K) \le T_n(K^{3\varepsilon}) + 3\varepsilon. \tag{1.7}$$

Thus, $\mathfrak{L}(X_n, X; \mathcal{M}(K_0)) \le 3\varepsilon$. Letting ε be sufficiently small proves the necessity. □

Corollary 1.5 *The random closed set X_n converges weakly to a compact random set X if and only if*

$$\mathfrak{L}(X_n, X) \to 0 \quad as \quad n \to \infty.$$

PROOF. *Sufficiency* immediately follows from Theorem 1.4.

Necessity. Let K_n, $n \ge 1$, be an increasing sequence of compacts, such that $K_n \uparrow \mathbb{R}^d$ as $n \to \infty$. Then, for a certain n,

$$T(\mathbb{R}^d) - T(K_n) < \varepsilon.$$

It is easy to show that the compact $K' = K_n^{\delta}$ belongs to $\mathcal{S}$ for a certain $\delta > 0$. Then

$$T_n(\mathbb{R}^d) - T_n(K') < \varepsilon$$

for sufficiently large n. For each $K \subseteq K'$ the inequalities (1.6) and (1.7) hold. If $K \not\subseteq K'$, then

$$T_n(K) \le T_n(K \cap K') + \varepsilon \le T(K^{3\varepsilon}) + 4\varepsilon$$

and

$$T(K) \le T_n(K^{3\varepsilon}) + 4\varepsilon.$$

Thus, $\mathfrak{L}(X_n, X) \to 0$ as $n \to \infty$. □

The introduced metrics depend on the class $\mathcal{M} \subseteq \mathcal{K}$. It is of great importance to choose this class properly. Hereafter we suppose that

$$c\mathcal{M} = \{cK \colon K \in \mathcal{M}\} = \mathcal{M}$$

for all $c > 0$, and also $K^r \in \mathcal{M}$ for all $r > 0$ and K from $\mathcal{M}$. We then say that $\mathcal{M}$ is *standard.* In the sequel a standard class may be safely thought to be the class of all closed balls.

7.2 Some Inequalities between Probability Metrics.

Derive several inequalities between the introduced probability metrics on the space of random sets distributions. Recall that the classical inequality between uniform and Levy metrics involves concentration functions of random variables. First, introduce the same notion for random sets.

The *concentration function* of a random closed set X is defined as

$$Q(\varepsilon, X; \mathcal{M}) = \sup \{T_X(K^\varepsilon) - T_X(K) \colon K \in \mathcal{M}\}, \ \varepsilon > 0, \tag{2.1}$$

cf. Hengartner and Theodorescu (1973). Evidently, $Q(\varepsilon, X; \mathcal{M})$ coincides with the uniform distance between the distributions of X and X^ε, i.e. $Q(\varepsilon, X; \mathcal{M})$ is equal to $\mathfrak{r}(X, X^\varepsilon; \mathcal{M})$. Other examples of concentration functions can be obtained by replacing $\mathfrak{r}$ with other probability metrics. As above, $Q(\varepsilon, X)$ means $Q(\varepsilon, X; \mathcal{K})$.

The following theorem provides an inequality between uniform and Levy metrics.

Theorem 2.1 *If $L = \mathfrak{L}(X, Y; \mathcal{M})$, then*

$$L \leq \mathfrak{r}(X, Y; \mathcal{M}) \leq L + \min \{Q(L, X; \mathcal{M}), Q(L, Y; \mathcal{M})\}. \tag{2.2}$$

PROOF follows from the obvious inequalities $L \leq \mathfrak{r}(X, Y; \mathcal{M})$ and

$$\begin{aligned} T_X(K) - T_Y(K) &= T_X(K) - T_Y(K^\varepsilon) + T_Y(K^\varepsilon) - T_Y(K) \\ &\leq \mathfrak{L}(X, Y; \mathcal{M}) + Q(\varepsilon, Y; \mathcal{M}) \end{aligned}$$

for $\varepsilon \leq \mathfrak{L}(X, Y; \mathcal{M})$. □

Consider some properties of the concentration function (2.1).

Theorem 2.2 *Let X and Y be independent random closed sets. Then, for each $\varepsilon > 0$ and $\mathcal{M} \subseteq \mathcal{K}$:*

1. $Q(\varepsilon, X \cup Y; \mathcal{M}) \leq Q(\varepsilon, X; \mathcal{M}) + Q(\varepsilon, Y; \mathcal{M})$.
2. $Q(\varepsilon, X/c; \mathcal{M}) = Q(\varepsilon, X; c\mathcal{M})$, $c \neq 0$.
3. $Q(\varepsilon, X \oplus Y) \leq \min(Q(\varepsilon, X), Q(\varepsilon, Y))$.

PROOF. It follows from (1.5.1) that

$$\begin{aligned} Q(\varepsilon, X \cup Y; \mathcal{M}) &= \sup \Big\{ [T_X(K^\varepsilon) - T_X(K)]\,(1 - T_Y(K^\varepsilon)) \\ &\qquad + [T_Y(K^\varepsilon) - T_Y(K)]\,(1 - T_X(K)) \colon K \in \mathcal{M} \Big\} \\ &\leq Q(\varepsilon, X; \mathcal{M}) + Q(\varepsilon, Y; \mathcal{M}). \end{aligned}$$

The second statement is obvious.

Note that

$$Q(\varepsilon, X) = \sup \{T_X(F^\varepsilon) - T_X(F) \colon F \in \mathcal{F}\},$$

since any $F \in \mathcal{F}$ is approximated by an increasing sequence of compact sets. Then

$$\begin{aligned} Q(\varepsilon, X \oplus Y) &= \sup \{\mathbf{E}[T_X(K^\varepsilon \oplus \check{Y}) - T_X(K \oplus Y) \mid Y] : K \in \mathcal{K}\} \\ &\le \sup \{\mathbf{E}[Q(\varepsilon, X) \mid Y] : K \in \mathcal{K}\} \\ &= Q(\varepsilon, X), \end{aligned}$$

where $\mathbf{E}[\cdot|\cdot]$ stands for conditional expectation. □

Considered as a function of z the concentration function $Q(z, X; \mathcal{M})$ is monotone. Moreover, it is left-continuous in case $\mathcal{M} \subseteq \mathcal{S}_T$.

Let L_z be the Levy distance between the distribution functions $F_X(z) = Q(z, X; \mathcal{M})$ and $F_Y(z) = Q(z, Y; \mathcal{M})$, i.e.

$$L_z = \inf \{\varepsilon > 0 : F_X(z) \le F_Y(z + \varepsilon) + \varepsilon,\ F_Y(z) \le F_X(z + \varepsilon) + \varepsilon, z \ge 0\}.$$

Theorem 2.3 *Let the class $\mathcal{M}$ be standard. Then*

$$L_z \le 2L + \max \{Q(L, X; \mathcal{M}), Q(L, Y; \mathcal{M})\},$$

where $L = \mathfrak{L}(X, Y; \mathcal{M})$.

PROOF. If $L \le \delta$, then $K^{z+\delta} \in \mathcal{M}$ for each $K \in \mathcal{M}$, $z > 0$, and

$$\begin{aligned} T_X(K^{z+\delta}) &\le T_Y(K^{z+2\delta}) + \delta, \\ T_Y(K) &\le T_X(K^\delta) + \delta. \end{aligned}$$

Thus

$$T_X(K^{z+\delta}) - T_X(K^\delta) \le T_Y(K^{z+2\delta}) + \delta - (T_Y(K) - \delta).$$

It is obvious that

$$\begin{aligned} T_X(K^{z+\delta}) - T_X(K^\delta) &\ge T_X(K^z) - T_X(K) + T_X(K) - T_X(K^\delta) \\ &\ge T_X(K^z) - T_X(K) + Q(\delta, X; \mathcal{M}). \end{aligned}$$

Hence

$$T_X(K^z) - T_X(K) - Q(\delta, X; \mathcal{M}) \le T_Y(K^{z+2\delta}) - T_Y(K) + 2\delta,$$

i.e.

$$Q(z, X; \mathcal{M}) \le Q(z + 2\delta, Y; \mathcal{M}) + 2\delta + Q(z, X; \mathcal{M}), z \ge 0.$$

Thus

$$F_X(z) \le F_Y(z + 2\delta) + 2\delta + Q(\delta, X; \mathcal{M}).$$

The similar inequality is valid on replacing X with Y. □

Evaluate concentration functions and distances for several examples of random closed sets.

EXAMPLE 2.4 Let $X = \{\xi\}$ and $Y = \{\eta\}$ be single-point random sets. Then $\mathfrak{L}(X,Y)$ is the Levy-Prohorov distance between ξ and η, see Zolotarev (1986). The uniform distance between X and Y is equal to

$$\mathfrak{r}(X,Y;\mathcal{M}) = \sup\{|\mathbf{P}\{\xi \in K\} - \mathbf{P}\{\eta \in K\}|\colon K \in \mathcal{M}\}.$$

If $\mathcal{M} = \mathcal{K}$, then $\mathfrak{r}(X,Y)$ coincides with the total variation distance between the distributions of ξ and η. The concentration function of X is equal to

$$Q(\varepsilon,X;\mathcal{M}) = \sup\{\mathbf{P}\{\xi \in K^\varepsilon\} - \mathbf{P}\{\xi \in K\}\colon K \in \mathcal{M}\}.$$

EXAMPLE 2.5 Let $X = (-\infty,\xi]$ and $Y = (-\infty,\eta]$ be random subsets of $\mathbb{R}^1$, and let $\{\inf K\colon K \in \mathcal{M}\} = \mathbb{R}^1$ (i.e. the class $\mathcal{M}$ is sufficiently "rich"). Then $\mathfrak{r}(X,Y;\mathcal{M})$ coincides with the uniform distance between ξ and η. Furthermore, $\mathfrak{L}(X,Y;\mathcal{M})$ is equal to the Levy distance between these random variables. The concentration functions of X and Y are equal to the classical concentration functions of the corresponding random variables, see Hengartner and Theodorescu (1973).

EXAMPLE 2.6 Let X and Y be the Poisson point processes in $\mathbb{R}^d$ with the intensity measures Λ_X and Λ_Y respectively. Then

$$|T_X(K) - T_Y(K)| \le |\Lambda_X(K) - \Lambda_Y(K)|,$$

so that $\mathfrak{r}(X,Y)$ is not greater than the total variation distance between Λ_X and Λ_Y. Similarly,

$$Q(\varepsilon,X;\mathcal{M}) \le \sup\{\Lambda_X(K^\varepsilon) - \Lambda_X(K)\colon K \in \mathcal{M}\}.$$

If X and Y are stationary and have intensities λ_X and λ_Y, then

$$\mathfrak{r}(X,Y) = \left|\left(\frac{\lambda_X}{\lambda_Y}\right)^{\frac{\lambda_X}{\lambda_Y-\lambda_X}} - \left(\frac{\lambda_X}{\lambda_Y}\right)^{\frac{\lambda_Y}{\lambda_Y-\lambda_X}}\right|.$$

EXAMPLE 2.7 Let X and Y be the Boolean models in $\mathbb{R}^d$ generated by the stationary Poisson point processes with the intensities λ_X, λ_Y and the primary grains A_X, A_Y (see Example 3.2.2). The capacity functional of the Boolean model X is equal to

$$T_X(K) = 1 - \exp\{-\lambda_X \mathbf{E}[\mu(K \oplus \check{A}_X)]\},$$

where $\check{A}_X = \{-x\colon x \in A_X\}$. Let $\mathcal{M}_0$ be the class of all balls. Then, by stationarity,

$$\mathfrak{r}(X,Y;\mathcal{M}_0) = \sup\left\{|\exp\{\lambda_X \mathbf{E}\mu(A_X^r)\} - \exp\{\lambda_Y \mathbf{E}\mu(A_Y^r)\}|\colon r \ge 0\right\}.$$

If A_X and A_Y are almost surely convex, then the Steiner formula (see Stoyan et al., 1987, Matheron, 1975) yields

$$\mathbf{E}\mu(A_X^r) = \sum_{i=0}^{d} \binom{d}{i} \mathbf{E}W_i(A_X) r^i,$$

where $W_i(A_X)$, $0 \le i \le d$, are Minkowski functionals of A_X. Hence, the distance $\mathfrak{r}(X,Y;\mathcal{M}_0)$ can be expressed in terms of the intensities λ_X, λ_Y and the expected values of the Minkowski functionals of the grains A_X, A_Y. For example, in $\mathbb{R}^2$ the functional W_0 is the area, W_1 is the perimeter length, and $W_2 = \pi$. Hence

$$\mathfrak{r}(X,Y;\mathcal{M}_0) = \sup\Big\{\big|\exp\{\lambda_X(S_X+2P_Xr+\pi r^2)\} - \exp\{\lambda_Y(S_Y+2P_Yr+\pi r^2)\}\big|:\ r\geq 0\Big\},$$

where S_X, P_X and S_Y, P_Y are the mean values of the area and the perimeter length of A_X, A_Y respectively. For example, if $\lambda_X=\lambda_Y=\lambda$ and $P_X=P_Y$, then

$$\mathfrak{r}(X,Y;\mathcal{M}_0) = |\exp\{-\lambda S_X\} - \exp\{-\lambda S_Y\}|.$$

Now evaluate distances between U-stable random closed sets defined in Chapter 3. It was proven that the capacity functional of a U-stable random set X is equal to

$$T_X(K) = 1-\exp\{\Psi_X(K)\}, \tag{2.3}$$

where Ψ_X is a homogeneous Choquet capacity such that

$$\Psi_X(sK) = s^\alpha \Psi_X(K)$$

for $s>0$ and all K, which misses the set of fixed points of X. The value α is said to be the index of X.

Hereafter

$$\mathcal{M}_0 = \{B_r(x):\ r\geq 0, x\in\mathbb{R}^d\}$$

designates the class of all balls and

$$\mathcal{M}_K = \{sK:\ s\geq 0\}$$

is the class of all scale transformations of the compact set K.

If X and Y are U-stable with the same index α then

$$\mathfrak{r}(X,Y;\mathcal{M}_K) = \sup\Big\{|\exp\{-s^\alpha\Psi_X(K)\} - \exp\{-s^\alpha\Psi_Y(K)\}|:\ s\geq 0\Big\}.$$

Hence

$$\mathfrak{r}(X,Y;\mathcal{M}_K) = h\left(\frac{\Psi_X(K)}{\Psi_Y(K)}\right), \tag{2.4}$$

where

$$h(x) = |x^c - x^{cx}|,\quad c = 1/(1-x).$$

Similarly,

$$Q(\varepsilon,X;\mathcal{M}_K) = h\left(\frac{\Psi_X(K^\varepsilon)}{\Psi_Y(K)}\right), \tag{2.5}$$

It follows from (2.4) and (2.5) that

$$\mathfrak{r}(X,Y;\mathcal{M}_0) = \sup\left\{h\left(\frac{\Psi_X(B_1(a))}{\Psi_Y(B_1(a))}\right):\ a\in\mathbb{R}^d\right\}, \tag{2.6}$$

$$Q(\varepsilon,X;\mathcal{M}_0) = \sup\left\{h\left(\frac{\Psi_X(B_{1+\varepsilon}(a))}{\Psi_X(B_1(a))}\right):\ a\in\mathbb{R}^d\right\}.$$

If the random sets X and Y are *stationary*, then

$$\begin{aligned} \mathfrak{r}(X,Y;\mathcal{M}_0) &= \mathfrak{r}(X,Y;\mathcal{M}_{B_1(0)}) \\ &= h\left(\frac{\Psi_X(B_1(0))}{\Psi_Y(B_1(0))}\right), \qquad (2.7) \\ Q(\varepsilon,X;\mathcal{M}_0) &= h((1+\varepsilon)^\alpha). \qquad (2.8) \end{aligned}$$

For example, if X and Y are stationary Poisson point processes, then

$$\mathfrak{r}(X,Y;\mathcal{M}_0) = \mathfrak{r}(X,Y) = h\left(\frac{\lambda_X}{\lambda_Y}\right).$$

7.3 Ideal Metrics for Random Closed Sets.

It was shown in Zolotarev (1979, 1986) that so-called ideal metrics play a significant role in the study of limit theorems.

The probability metric $\mathfrak{m}$ is said to be *ideal* if $\mathfrak{m}$ is homogeneous and regular. Namely, $\mathfrak{m}$ is *homogeneous* of degree γ if

$$\mathfrak{m}(cX,cY) = |c|^\gamma \mathfrak{m}(X,Y), \qquad (3.1)$$

whatever $c \neq 0$ may be. The metric $\mathfrak{m}$ is said to be *regular* with respect to unions if

$$\mathfrak{m}(X\cup Z, Y\cup Z) \leq \mathfrak{m}(X,Y) \qquad (3.2)$$

for any random set Z independent of X and Y. Regular metrics with respect to the Minkowski addition are defined similarly.

Hereafter suppose that the class $\mathcal{M}$ is *standard* if otherwise is not stated. Then the uniform metric $\mathfrak{r}(X,Y;\mathcal{M})$ is ideal of zero degree. Indeed,

$$\mathfrak{r}(cX,cY;\mathcal{M}) = \mathfrak{r}(X,Y;\mathcal{M}/c) = \mathfrak{r}(X,Y;\mathcal{M}), \qquad (3.3)$$

and

$$\begin{aligned} \mathfrak{r}(X\cup Z, Y\cup Z;\mathcal{M}) &= \sup\{|T_{X\cup Z}(K) - T_{Y\cup Z}(K)|:\ K\in\mathcal{M}\} \\ &= \sup\{|T_X(K) + T_Z(K) - T_X(K)T_Z(K) - T_Y(K) \\ &\qquad -T_Z(K) + T_Z(K)T_Y(K)| :\ K\in\mathcal{M}\} \\ &= \sup\{|(T_X(K) - T_Y(K))|(1-T_Z(K)):\ K\in\mathcal{M}\} \\ &\leq \mathfrak{r}(X,Y;\mathcal{M}). \end{aligned}$$

If $\mathcal{M} = \mathcal{K}$, then the metric $\mathfrak{r}(X,Y) = \mathfrak{r}(X,Y;\mathcal{K})$ is regular with respect to the Minkowski addition. Indeed,

$$\begin{aligned} \mathfrak{r}(X\oplus Z, Y\oplus Z) &= \sup\left\{\left|\mathbf{E}\left[T_X(K\oplus\check{Z}) - T_Y(K\oplus\check{Z})\mid Z\right]\right|:\ K\in\mathcal{K}\right\} \\ &\leq \mathfrak{r}(X,Y). \end{aligned}$$

The Levy metric $\mathfrak{L}(X,Y;\mathcal{M})$ is also regular with respect to unions and with respect to the Minkowski addition in case $\mathcal{M}=\mathcal{K}$. For example, $\mathfrak{L}(X,Y;\mathcal{M})\leq\varepsilon$ yields

$$\begin{aligned}
T_{X\cup Z}(K) &= T_X(K)+T_Z(K)-T_X(K)T_Z(K)\\
&= T_X(K)(1-T_Z(K))+T_Z(K)\\
&\leq T_Y(K^\varepsilon)-T_Y(K^\varepsilon)T_Z(K)+T_Z(K)+\varepsilon\\
&\leq T_Y(K^\varepsilon)+\varepsilon+T_Z(K^\varepsilon)(1-T_Y(K^\varepsilon))\\
&= T_{Y\cup Z}(K^\varepsilon)+\varepsilon,
\end{aligned}$$

so that $\mathfrak{L}(X\cup Z,Y\cup Z;\mathcal{M})\leq\varepsilon$ too.

It follows from (3.1) and (3.3) that $\mathfrak{r}$ and $\mathfrak{L}$ are ideal metrics of *zero* degree. Nevertheless, ideal metrics of a *positive* degree are preferable. To apply the probability metrics method to limit theorems for unions of random sets we have to find out an ideal metric of positive degree γ. Such a metric can be constructed by generalizing the uniform metric $\mathfrak{r}$. Put

$$\mathfrak{r}_\Phi(X,Y;\mathcal{M})=\sup\left\{\Phi(K)|T_X(K)-T_Y(K)|:\ K\in\mathcal{M}\right\}, \tag{3.4}$$

where $\Phi:\mathcal{K}\longrightarrow[0,\infty)$ is a non-negative increasing and homogeneous functional of degree $\gamma>0$, i.e.

$$\Phi(sK)=s^\gamma\Phi(K), \tag{3.5}$$

whatever positive s and K from $\mathcal{K}$ may be.

We may put, for example, $\Phi(K)=(\mu(K))^{\gamma/d}$ or $\Phi(K)=(C(K))^\gamma$, where μ is the Lebesgue measure, C is the Newton capacity, see Landkof (1966), Matheron (1975).

Hereafter suppose that Φ is chosen in such a way that for each $K\in\mathcal{K}$

$$\Phi(K^\delta)\rightarrow\Phi(K^\varepsilon)\ \text{ as }\ \delta\rightarrow\varepsilon>0.$$

Clearly, the metric $\mathfrak{r}_\Phi$ is an ideal metric of degree γ with respect to unions.

Derive an inequality between $\mathfrak{r}_\Phi$ and the Levy metric $\mathfrak{L}$. For any $K_0\in\mathcal{K}$ and $\delta>0$ introduce the family of compacts by means of

$$\mathcal{U}_\delta(K_0)=\left\{K\in\mathcal{K}:\ K_0\subseteq K\subseteq K_0^\delta\right\}.$$

Lemma 3.1 *The value $\mathfrak{L}(X,Y;\mathcal{M})$ is equal to the supremum $\hat{L}$ of all positive δ such that, for a certain compact $K_0\in\mathcal{M}$ and any K_1, K_2 belonging to $\mathcal{U}_\delta(K_0)\cap\mathcal{M}$,*

$$|T_X(K_1)-T_Y(K_2)|\geq\delta.$$

PROOF. Let $\hat{L}<\delta$, and let $K\in\mathcal{M}$ be specified. Then

$$|T_X(K_1)-T_Y(K_2)|\leq\delta$$

for some K_1, K_2 from $\mathcal{U}_\delta(K_0)\cap\mathcal{M}$. Hence

$$T_X(K)\leq T_X(K_1)\leq T_Y(K_2)+\delta\leq T_Y(K^\delta)+\delta.$$

Similarly,

$$T_Y(K) \leq T_X(K^\delta) + \delta.$$

Thus, $\mathfrak{L}(X, Y; \mathcal{M}) \leq \delta$, so that $\mathfrak{L}(X, Y; \mathcal{M}) \leq \hat{L}$.

Let $\mathfrak{L}(X, Y; \mathcal{M}) < \delta$. For sufficiently small $\varepsilon > 0$ there exists a compact K_0 such that

$$T_X(K_0) \geq T_Y(K_0^{\delta-\varepsilon}) + \delta - \varepsilon$$

or

$$T_Y(K_0) \geq T_X(K_0^{\delta-\varepsilon}) + \delta - \varepsilon.$$

Then, for each K from $\mathcal{U}_{\delta-\varepsilon}(K_0)$,

$$T_X(K) \geq T_X(K_0) \geq T_Y(K_0^{\delta-\varepsilon}) + \delta - \varepsilon \geq T_Y(K) + \delta - \varepsilon$$

or

$$T_Y(K) \geq T_X(K) + \delta - \varepsilon.$$

Hence $\hat{L} \leq \delta\varepsilon$ and, therefore, $\hat{L} \leq \mathfrak{L}(X, Y; \mathcal{M})$. □

It is well-known that the Levy distance between distribution functions is equal to the side of the maximal square inscribed between the graphs of the functions in question. Lemma 3.1 generalized this property for the Levy distance between capacities. The family $\mathcal{U}_\delta(K_0)$ plays the role of the side of the "square" incribed between the graphs of capacities, see also Baddeley (1991).

Theorem 3.2 *If* $L = \mathfrak{L}(X, Y; \mathcal{M})$, *then*

$$\mathfrak{r}_\Phi(X, Y; \mathcal{M}) \geq L^{1+\gamma} \inf_{x \in \mathbb{R}^d} \Phi(B_1(x)). \tag{3.6}$$

PROOF. If $\delta < L$, then Lemma 3.1 yields

$$\left|T_X(K_0^\delta) - T_Y(K_0^\delta)\right| \geq \delta$$

for a certain compact $K_0 \in \mathcal{M}$. Hence

$$\begin{aligned} \mathfrak{r}_\Phi(X, Y; \mathcal{M}) &\geq \Phi(K_0^\delta)\left|T_X(K_0^\delta) - T_Y(K_0^\delta)\right| \\ &\geq \delta\Phi(K_0^\delta). \end{aligned}$$

It follows from (3.5) that

$$\begin{aligned} \mathfrak{r}_\Phi(X, Y; \mathcal{M}) &\geq L \inf_{K \in \mathcal{M}} \Phi(K^L) \\ &\geq L \inf_{x \in \mathbb{R}^d} \Phi(B_L(x)) \\ &= L^{1+\gamma} \inf_{x \in \mathbb{R}^d} \Phi(B_1(x)). \quad \square \end{aligned}$$

Corollary 3.3 *If the functional* Φ *is shift-invariant, then*

$$\mathfrak{r}_\Phi(X, Y; \mathcal{M}) \geq \mathfrak{L}(X, Y; \mathcal{M})^{1+\gamma}\Phi(B_1(0)). \tag{3.7}$$

From (2.2) and (3.7) we obtain the following inequality between $\mathfrak{r}_\Phi$ and $\mathfrak{r}$:

$$\mathfrak{r}(X,Y;\mathcal{M}) \leq q + \min\left(Q(q,X;\mathcal{M}), Q(q,Y;\mathcal{M})\right), \tag{3.8}$$

where

$$q = \left(\frac{\mathfrak{r}_\Phi(X,Y;\mathcal{M})}{\Phi(B_1(0))}\right)^{\gamma_1} \tag{3.9}$$

and

$$\gamma_1 = \frac{1}{1+\gamma}. \tag{3.10}$$

It follows from (2.8) that

$$\mathfrak{r}(X,Y;\mathcal{M}) \leq q + h((1+q)^d), \tag{3.11}$$

in case X is stationary and U-stable.

If the functional Φ is decreasing and satisfies (3.5) with $\gamma < 0$, then

$$\mathfrak{r}_\Phi(X,Y;\mathcal{M}(K_0)) \geq L\Phi(K_0^L)$$

for all $K_0 \in \mathcal{K}$.

The statement of Theorem 3.2 is valid for any class $\mathcal{M}$ such that K^δ belongs to $\mathcal{M}$ for all $\delta > 0$ and $K \in \mathcal{M}$. However, this condition is too restrictive. Consider a generalization of Theorem 3.2 for the shift-invariant functional Φ.

Theorem 3.4 *Let $\mathcal{M} \subseteq \mathcal{M}_1 \subseteq \mathcal{K}$, and let $K^r \in \mathcal{M}_1$ for each K belonging to $\mathcal{M}$, $r \in [0,\varepsilon]$ and a certain $\varepsilon > 0$. Then*

$$\mathfrak{r}_\Phi(X,Y;\mathcal{M}_1) \geq \min\left(\mathfrak{L}(X,Y;\mathcal{M}),\varepsilon\right)^{1+\gamma} \Phi(B_1(0)).$$

PROOF. Let $\delta < \mathfrak{L}(X,Y;\mathcal{M})$. Lemma 3.1 yields

$$\left|T_X(K_0^\delta) - T_Y(K_0^\delta)\right| \geq \delta$$

in case $\delta \leq \varepsilon$ or

$$\left|T_X(K_0^\varepsilon) - T_Y(K_0^\varepsilon)\right| \geq \delta$$

when $\delta \geq \varepsilon$.

The definition of the metric $\mathfrak{r}_\Phi$ yields

$$\mathfrak{r}_\Phi(X,Y;\mathcal{M}_1) \geq \Phi(K_0^\delta)\delta$$

for $\delta < \varepsilon$, since $K_0^\delta \in \mathcal{M}_1$. Otherwise

$$\mathfrak{r}_\Phi(X,Y;\mathcal{M}_1) \geq \Phi(K_0^\varepsilon)\delta \geq \Phi(K_0^\varepsilon)\varepsilon.$$

The proof can be finished similar to the proof of Theorem 3.2. □

Corollary 3.5 *Let*

$$\mathcal{M}(a) = \left\{B_r(x)\colon r > 0, x \in \mathbb{R}^d, \|x\| - r \geq a\right\}$$

for a certain positive a. Then, for each $a_0 < a$,

$$\mathfrak{r}_\Phi(X,Y;\mathcal{M}(a_0)) \geq \min\left\{\mathfrak{L}(X,Y;\mathcal{M}(a)), a - a_0\right\}^{1+\gamma} \Phi(B_1(0)).$$

Note that $\mathcal{M}(a_0) \supset \mathcal{M}(a)$ when $a_0 < a$.

7.4 Applications to Limit Theorems for Unions.

In this section previously obtained results are applied to the study of limit theorems for unions of random closed sets. Let $A_1, A_2, \ldots$ be iid copies of a certain random closed set A, and let

$$Y_n = n^{1/\alpha}(A_1 \cup \cdots \cup A_n).$$

If Y_n converges weakly to a certain non-trivial random set X, then the limiting random set is U-stable with index α.

For any $F \subset \mathbb{R}^d$ denote its "inverse" set by

$$F^* = \{x\|x\|^{-2}\colon\, x \in F\}, \tag{4.1}$$

see also Section 3.1.

Theorem 4.1 *Let X be a U-stable random set with the negative index α, and let Φ be a homogeneous increasing functional, which satisfies (3.5) for $\gamma > -\alpha$. If the distance $\mathfrak{r}_\Phi(A_1, X; \mathcal{M})$ is finite, then*

$$\mathfrak{r}_\Phi(Y_n, X; \mathcal{M}) \le n^{1+\frac{\gamma}{\alpha}} \mathfrak{r}_\Phi(A_1, X; \mathcal{M}). \tag{4.2}$$

If X is U-stable with index $\alpha > 0$ and $\mathfrak{r}_\Phi(A_1^*, X^*; \mathcal{M}) < \infty$, $\gamma > \alpha$, then

$$\mathfrak{r}_\Phi(Y_n, X; \mathcal{M}) \le n^{1-\frac{\gamma}{\alpha}} \mathfrak{r}_\Phi(A_1^*, X^*; \mathcal{M}). \tag{4.3}$$

PROOF is similar to the proof of the corresponding result for random variables, see Zolotarev (1986). Since X is U-stable,

$$X \stackrel{d}{\sim} n^{1/\alpha}(X_1 \cup \cdots \cup X_n),$$

for iid random sets $X_1, \ldots, X_n$ equivalent to X. It follows from (3.1), (3.2) that

$$\begin{aligned}
\mathfrak{r}_\Phi(Y_n, X; \mathcal{M}) &= \mathfrak{r}_\Phi\left(n^{1/\alpha}(A_1 \cup \cdots \cup A_n), n^{1/\alpha}(X_1 \cup \cdots \cup X_n); \mathcal{M}\right) \\
&\le n^{\gamma/\alpha} \sum_{k=1}^{n} \mathfrak{r}_\Phi(A_k, X_k; \mathcal{M}) \\
&= n^{1+\frac{\gamma}{\alpha}} \mathfrak{r}_\Phi(A_1, X; \mathcal{M}).
\end{aligned}$$

Thus, (4.2) is valid.

It is easy to show that $(cF)^* = c^{-1}F^*$. Hence

$$Y_n^* = n^{-1/\alpha}(A_1^* \cup \cdots \cup A_n^*),$$

so that (4.3) immediately follows from (4.2). □

Note that Theorem 4.1 is valid for all classes $\mathcal{M} \subseteq \mathcal{K}$, such that $c\mathcal{M} = \mathcal{M}$ for all *sufficiently large c.*

If the class $\mathcal{M}$ is *standard*, then Theorem 4.1 and (3.8) yield

$$\mathfrak{r}(Y_n, X; \mathcal{M}) \le q + Q(q, X; \dot{\mathcal{M}}), \tag{4.4}$$

where

$$q = \left(n^{1+\frac{\gamma}{\alpha}} \frac{\mathfrak{r}_\Phi(A_1, X; \mathcal{M})}{\Phi(B_1(0))} \right)^{1/(1+\gamma)}.$$

The inequality (4.4) provides an estimate for the speed of convergence of capacity functionals in the limit theorem for unions of random sets.

Naturally, the pointwise and even the uniform convergence of capacity functionals on $\mathcal{M}$ follow from (4.4). In particular, let $\mathcal{M} = \mathcal{M}_K$, and let $\Phi(K) = (\mu(K))^{\gamma/d}$. Then

$$\mathfrak{r}(Y_n, X; \mathcal{M}_K) \leq q + h\left(\frac{\Psi_X(K^q)}{\Psi_X(K)}\right),$$

where

$$\begin{aligned} q &= n^\beta \left(\frac{\mathfrak{r}_\Phi(A_1, X; \mathcal{M}_K)}{b_d^{\gamma/d}} \right)^{1/(1+\gamma)}, \\ \beta &= \frac{\alpha+\gamma}{\alpha(1+\gamma)}, \end{aligned}$$

b_d is the volume of the unit ball in $\mathbb{R}^d$.

Let us estimate the distance $\mathfrak{r}_\Phi(A_1, X; \mathcal{M}_K)$ in the scheme of Theorem 4.4.5.

Let $A_1 = M(\xi)$ be defined in Theorem 4.4.5. Here ξ is a random vector in $\mathbb{R}^m$ having the positive regularly varying density f, and $M: \mathbb{R}^m \longrightarrow \mathcal{K}$ is a homogeneous multifunction with values in the class $\mathcal{K}$ of compacts in $\mathbb{R}^d$ such that $M(su) = s^\eta M(u)$ for each $s > 0$, $u \in \mathbb{R}^m$ and a certain $\eta > 0$.

Suppose that $\operatorname{ind} f = \alpha - m$ for some $\alpha < 0$, and $f(u) = \phi(u)L(u)$, where ϕ is homogeneous and L is slowly varying. These conditions have been already used in Theorem 4.4.5. In addition, suppose that

$$L(su) \to 1 \text{ as } s \to \infty.$$

The functional $\Phi(K)$ from (3.4) is defined as

$$\Phi(K) = (\mu(K))^{\gamma/d},$$

where $\gamma > 0$ and μ is the Lebesgue measure.

Consider the U-stable random set X with the capacity functional

$$T(K) = 1 - \exp\{\Lambda(\mathcal{L}_K)\}, \tag{4.5}$$

where

$$\Lambda(\mathcal{L}_K) = \int_{\mathcal{L}_K} \phi(u) du, \tag{4.6}$$

$$\mathcal{L}_K = \{u \in \mathbb{R}^d \colon M(u) \cap K \neq \emptyset\}. \tag{4.7}$$

Note that $\mathcal{L}_K$ is compact in case $0 \notin K$ and the index of the U-stable random set X is equal to α/η, since $\Lambda(\mathcal{L}_{xK}) = x^{\alpha/\eta}\Lambda(\mathcal{L}_K)$.

Theorem 4.2 *Let $\beta = -\alpha/\eta$, and let $\beta < \gamma < 2\beta$. Suppose that for $K \in \mathcal{K}$, $0 \notin K$*

$$\sup_{u \in \mathcal{L}_K} (L(us) - 1) = o(s^{\eta(\beta-\gamma)}) \quad as \quad s \to \infty. \tag{4.8}$$

Then $\mathfrak{r}_\Phi(A_1, X; \mathcal{M}_K)$ is finite. Moreover,

$$\mathfrak{r}_\Phi(A_1, X; \mathcal{M}_K) \le \mu(K)^{\gamma/d} \left(C(\gamma/\beta)\Lambda(\mathcal{L}_K)^{\gamma/\beta} + \kappa(\mathcal{L}_K)\right),$$

where

$$\begin{aligned} C(x) &= \sup_{y \ge 0} y^{-x}(y - 1 + e^{-y}), \\ \kappa(F) &= \sup_{s \ge 0} s^{\gamma-\beta} \left| \int_F \phi(u) \left(L(us^{1/\eta}) - 1 \right) du \right|. \end{aligned}$$

PROOF. It is evident that

$$\begin{aligned} \mathfrak{r}_\Phi(A_1, X; \mathcal{M}_K) &= \sup \left\{ \mu(K)^{\gamma/d} s^\gamma |\mathbf{P}\{\xi \in \mathcal{L}_{sK}\} - (1 - \exp\{-\Lambda(\mathcal{L}_{sK})\})| : s \ge 0 \right\} \\ &= \mu(K)^{\gamma/d} \sup \left\{ s^\gamma \left| s^{-\beta} \int_{\mathcal{L}_K} \phi(u) L(us^{1/\eta}) du \right.\right. \\ &\qquad \left.\left. - \left(1 - \exp\left\{-s^{-\beta}\Lambda(\mathcal{L}_K)\right\}\right) \right| : s \ge 0 \right\} \\ &= \mu(K)^{\gamma/d} \sup \left\{ t^{-\gamma/\beta} \left| t \int_{\mathcal{L}_K} \phi(u) L(ut^{1/\alpha}) du \right.\right. \\ &\qquad \left.\left. -1 + \exp\left\{-t\Lambda(\mathcal{L}_K)\right\} \right| : t \ge 0 \right\}. \end{aligned} \tag{4.9}$$

In view of $\beta < \gamma < 2\beta$, we get

$$\begin{aligned} & t^{-\gamma/\beta} \left(t \int_{\mathcal{L}_K} \phi(u) L(t^{1/\alpha} u) du - 1 + \exp\left\{-t\Lambda(\mathcal{L}_K)\right\} \right) \\ & \qquad \sim t^{-\gamma/\beta} t L(t^{1/\alpha} e) \int_{\mathcal{L}_K} \phi(u) du \to 0 \text{ as } t \to \infty. \end{aligned}$$

It follows from (4.8) that

$$\begin{aligned} & t^{-\gamma/\beta} \left(t \int_{\mathcal{L}_K} \phi(u) L(t^{1/\alpha} u) du - 1 + \exp\left\{-t\Lambda(\mathcal{L}_K)\right\} \right) \\ & \qquad \sim t^{-\gamma/\beta} \left(t \int_{\mathcal{L}_K} \phi(u) L(t^{1/\alpha} u) du - t\Lambda(\mathcal{L}_K) + \mathcal{O}(t^2) \right) \\ & \qquad \sim t^{-\gamma/\beta} \left(t \int_{\mathcal{L}_K} \phi(u) \left(L(t^{1/\alpha} u) - 1 \right) du + \mathcal{O}(t^2) \right) \to 0 \text{ as } t \to 0. \end{aligned}$$

Thus, the distance $\mathfrak{r}_\Phi(A_1, X; \mathcal{M}_K)$ is finite.

It follows from (4.9) that

$$\begin{aligned}
\mathfrak{r}_{\Phi}(A_1, X; \mathcal{M}_K) &\leq \mu(K)^{\gamma/d}\Big(\sup\left\{t^{-\gamma/\beta}|t\Lambda(\mathcal{L}_K) - 1 + \exp\{-\Lambda(\mathcal{L}_K)\}|\colon t \geq 0\right\} \\
&\qquad + \sup\left\{t^{-\gamma/\beta+1}\left|\int_{\mathcal{L}_K} \phi(u)\left(L(t^{1/\alpha}u) - 1\right) du\right|\colon t \geq 0\right\}\Big) \\
&= \mu(K)^{\gamma/d}\Big(\Lambda(\mathcal{L}_K)^{\gamma/\beta} \sup\left\{y^{-\gamma/\beta}(y - 1 + e^{-y})\colon y \geq 0\right\} \\
&\qquad + \sup\left\{y^{\gamma-\beta}\left|\int_{\mathcal{L}_K} \phi(u)\left(L(y^{1/\eta}u) - 1\right) du\right|\colon y \geq 0\right\}\Big) \\
&= \mu(K)^{\gamma/d}\left(C(\gamma/\beta)\Lambda(\mathcal{L}_K)^{\gamma/\beta} + \kappa(\mathcal{L}_K)\right). \quad \square
\end{aligned}$$

NOTE. If

$$|L(us) - 1| \leq \min\left(1, \phi_1(u)s^{\Delta\eta}\right)$$

for $\Delta < \beta - \gamma$, then

$$\kappa(\mathcal{L}_K) \geq \Lambda(\mathcal{L}_K)^{(\gamma-\beta)/\Delta+1}\left(\int_{\mathcal{L}_K} \phi(u)\phi_1(u)du\right)^{-(\gamma-\beta)/\Delta}.$$

We can apply Theorem 4.1 to the metric $\mathfrak{r}_{\Phi}(A_1, X; \mathcal{M}_K)$ and obtain an estimate for $\mathfrak{r}_{\Phi}(Y_n, X; \mathcal{M}_K)$, which, in fact, represents the speed of convergence with respect to this metric. Nevertheless, these estimates *cannot* be reformulated in terms of the uniform metric $\mathfrak{r}(Y_n, X; \mathcal{M}_K)$, since Theorem 3.2 is no longer applicable to the class $\mathcal{M}_K$.

Recall that in Theorem 3.2 we assumed that $K^\delta \in \mathcal{M}$ for each $K \in \mathcal{M}$ and $\delta > 0$. Thus, we have to consider, e.g., the distance $\mathfrak{r}_{\Phi}(A_1, X; \mathcal{M}_0)$, where $\mathcal{M}_0$ is the class of all balls. However, for this class $\mathcal{M}_0$ the distance $\mathfrak{r}_{\Phi}(A_1, X; \mathcal{M}_0)$ is infinite, since the origin is a fixed point of the random set X. To obtain essential estimates in this case we make use of Theorem 3.4 and Corollary 3.5.

Let us estimate the distance $\mathfrak{r}_{\Phi}$ for the class $\mathcal{M}_0(a, R)$ defined as

$$\mathcal{M}_0(a, R) = \left\{B_r(x)\colon r > 0, x \in \mathbb{R}^d, \|x\| - r \geq a, B_r(x) \subseteq B_R(0)\right\}$$

for some $R > a > 0$. Unfortunately, the metric $\mathfrak{r}_{\Phi}(A_1, X; \mathcal{M}_0(a, R))$ is no longer homogeneous, since $c\mathcal{M}_0(a, R) \neq \mathcal{M}_0(a, R)$ for an arbitrary c.

To overcome this difficulty note that in Theorem 4.1 we did not use the full strength of the homogeneous property (3.3). It is sufficient to ensure

$$\mathfrak{m}(cX, cY) \leq \mathfrak{m}(X, Y) \tag{4.10}$$

for all sufficiently small c. Then the metric $\mathfrak{m}$ is said to be *semi-homogeneous*.

It is easy to show that the metric $\mathfrak{r}_\Phi(A_1, X; \mathcal{M})$ is semi-homogeneous if, for a certain $c_0 \geq 0$,

$$c\mathcal{M} \subseteq \mathcal{M} \text{ for all } c \geq c_0. \tag{4.11}$$

Introduce the class $\mathcal{M}(a, R)$ by

$$\mathcal{M}(a, R) = \{cK\colon K \in \mathcal{M}_0(a, R), c \geq 1\}. \tag{4.12}$$

Then $c\mathcal{M}(a, R) \subset \mathcal{M}(a, R)$ for all $c \geq 1$, whence the corresponding metric $\mathfrak{r}_\Phi$ is semi-homogeneous. Evidently,

$$\mathfrak{r}_\Phi(A_1, X; \mathcal{M}(a, R)) = \sup\{\mathfrak{r}_\Phi(A_1, X; \mathcal{M}_K)\colon K \in \mathcal{M}(a, R)\}.$$

Theorem 4.2 yields

$$\begin{aligned}
&\mathfrak{r}_\Phi(A_1, X; \mathcal{M}(a, R)) \leq \\
&\leq \sup\Big\{\mu(B_r(x))^{\gamma/d}\left[C(\gamma/\beta)\Lambda(\mathcal{L}_{B_r(x)})^{\gamma/\beta} + \kappa(\mathcal{L}_{B_r(x)})\right]: \\
&\qquad r > 0, x \in \mathbb{R}^d, B_r(x) \cap B_a(0) = \emptyset, B_r(x) \subset B_R(0)\Big\} \\
&= \sup\Big\{r^\gamma b_d^{\gamma/d}\left[C(\gamma/\beta)\Lambda(\mathcal{L}_{B_r(x)})^{\gamma/\beta} + \kappa(\mathcal{L}_{B_r(x)})\right]: \\
&\qquad 0 \leq r \leq (R-a)/2, a + r \leq \|x\| \leq R - r\Big\} \\
&= b_d^{\gamma/d} \sup\Big\{C(\gamma/\beta)\Lambda(\mathcal{L}_{B_1(x/r)})^{\gamma/\beta} + \kappa(\mathcal{L}_{B_1(x/r)}): \\
&\qquad 0 \leq r \leq (R-a)/2, a + r \leq \|x\| \leq R - r\Big\} \\
&\leq b_d^{\gamma/d} \sup\Big\{C(\gamma/\beta)\Lambda(\mathcal{L}_{B_1(v)})^{\gamma/\beta} + \kappa(\mathcal{L}_{B_1(v)}): \\
&\qquad 0 \leq r \leq (R-a)/2, a/r + 1 \leq \|v\|\Big\},
\end{aligned}$$

where b_d is the volume of the unit ball in $\mathbb{R}^d$. Finally,

$$\begin{aligned}
&\mathfrak{r}_\Phi(A_1, X; \mathcal{M}(a, R)) \leq \\
&\leq b_d^{\gamma/d} \sup\left\{C(\gamma/\beta)\Lambda(\mathcal{L}_{B_1(v)})^{\gamma/\beta} + \kappa(\mathcal{L}_{B_1(v)})\colon \|v\| \geq 1 + \frac{2a}{R-a}\right\}.
\end{aligned}$$

Consider the particular case $M(u) = \{u\}$, i.e. $M(u)$ is a single-point-valued function (see Theorem 4.4.1). Then

$$\mathcal{L}_{B_r(v)} = B_1(v),$$

and $\beta = -\alpha$. Suppose also that

$$|L(u_1 s) - 1| \leq |L(u_2 s) - 1|$$

if $\|u_1\| > \|u_2\|$. Then

$$\mathfrak{r}_\Phi(A_1, X; \mathcal{M}(a,R)) \le b_d^{\gamma/d} \sup\left\{C(\gamma/\beta)\Lambda(B_1(qw))^{\gamma/\beta} + \kappa(B_1(qw)) \colon \|w\| = 1\right\},$$

where $q = 1 + 2a/(R-a)$. If the distribution of ξ is spherically symmetric, then, for a certain $e \in \mathbb{S}^{d-1}$,

$$\mathfrak{r}_\Phi(A_1, X; \mathcal{M}(a,R)) \le b_d^{\gamma/d}\Big(C(\gamma/\beta)\Lambda(B_1(qe))^{\gamma/\beta} + \kappa(B_1(qe))\Big).$$

If, additionally, $L(u) = 1$ whenever $u \notin B_{q-1}(0)$ and $\eta = 1$, then

$$\mathfrak{r}_\Phi(A_1, X; \mathcal{M}(a,R)) \le b_d^{\gamma/d}\left\{C(-\gamma/\alpha)\Lambda(B_1(qe))^{-\gamma/\alpha}\right\}. \tag{4.13}$$

EXAMPLE **4.3** Let ξ have Cauchy distribution in $\mathbb{R}^1$. Then $\alpha = -1$, $\phi(u) = \pi u^{-2}$ and $L(u) = u^2/(1+u^2)$ for $u \ne 0$. Hence

$$\Lambda(B_1(q)) = \int_{q-1}^{q+1} \pi u^{-2} du = \pi\left((q-1)^{-1} - (q+1)^{-1}\right),$$

and

$$\begin{aligned}
\kappa(B_1(q)) &= \sup_{s\ge 0} s^{\gamma-1} \int_{q-1}^{q+1} \frac{\pi}{u^2(1+u^2)} du \\
&\le \pi \sup_{s\ge 0} \min\left(s^{\gamma-1}\int_{q-1}^{q+1} \pi u^{-2} du,\ s^{\gamma-3}\int_{q-1}^{q+1} \pi u^{-4} du\right) \\
&= \pi\left(\int_{q-1}^{q+1} \pi u^{-2} du\right)^{(\gamma+1)/2}\left(\int_{q-1}^{q+1} \pi u^{-4} du\right)^{(\gamma-1)/2} \\
&\le \pi 3^{(1-\gamma)/2}(q-1)^{-2\gamma+1} \\
&= \pi 3^{(1-\gamma)/2}\left(\frac{R-a}{2a}\right)^{2\gamma-1}.
\end{aligned}$$

Thus

$$\mathfrak{r}_\Phi(A_1, X; \mathcal{M}(a,R)) \le \pi 2^\gamma\left[C(-\gamma/\alpha)\left(\frac{R-a}{2a}\right) + 3^{(1-\gamma)/2}\left(\frac{R-a}{2a}\right)^{2\gamma-1}\right].$$

In view of (4.13), Theorem 4.1 allows to estimate the distance $\mathfrak{r}_\Phi(Y_n, X; \mathcal{M}(a,R))$. It follows from Theorem 3.4 that, for any $a_0 > a$ and $R_0 < R$,

$$\mathfrak{r}_\Phi(Y_n, X; \mathcal{M}(a,R)) \ge \min\left(\mathfrak{L}(Y_n, X; \mathcal{M}(a_0, R_0)),\ \varepsilon\right)^{1+\gamma}\Phi(B_1(0)),$$

where $\varepsilon = \min(a_0 - a, R - R_0)$.

Thus, for sufficiently large n such that

$$\mathfrak{r}_\Phi(Y_n, X; \mathcal{M}(a,R)) \le \varepsilon^{1+\gamma} b_d, \tag{4.14}$$

we get

$$\mathfrak{r}_\Phi(Y_n, X; \mathcal{M}(a,R)) \ge \mathfrak{L}(Y_n, X; \mathcal{M}(a_0, R_0))^{1+\gamma} b_d.$$

Hence, for sufficiently large n satisfying (4.14),

$$\mathfrak{L}(Y_n, X; \mathcal{M}(a_0, R_0)) \le \left(n^{1+\gamma/\beta}\frac{\mathfrak{r}_\Phi(A_1, X; \mathcal{M}(a, R))}{b_d}\right)^{1/(1+\gamma)}.$$

Finally, Theorem 2.1 allows to estimate the distance $\mathfrak{r}(Y_n, X; \mathcal{M}(a_0, R_0))$.

The estimate of $\mathfrak{r}_\Phi(A_1, X; \mathcal{M}(a, R))$ can be refined in the following way. Note that $\mathcal{M}(a, R)$ does not consist of the whole classes $\mathcal{M}_K$ for K belonging to $\mathcal{M}_0(a, R)$, but only compacts cK for $c \ge 1$ and K belonging to $\mathcal{M}_0(a, R)$ (cf. (4.12)). Denote

$$\mathcal{M}'_K = \{cK\colon c \ge 1\}.$$

Similarly to Theorem 4.2, we get

$$\begin{aligned}\mathfrak{r}_\Phi(A_1, X; \mathcal{M}'_K) \;&\le\; \mu(K)^{\gamma/d}\Bigg(\Lambda(\mathcal{L}_K)^{\gamma/\beta}\sup\{y^{-\gamma/\beta}(y-1+e^{-y})\colon 0 \le y \le \Lambda(\mathcal{L}_K)\} \\ &\quad + \sup\left\{y^{\gamma-\beta}\left|\int_{\mathcal{L}_K}\phi(u)\left(L(y^{1/\eta}u)-1\right)du\right|\colon y \ge 1\right\}\Bigg).\end{aligned}$$

In particular,

$$\begin{aligned}\mathfrak{r}_\Phi(A_1, X; \mathcal{M}'_{B_r(x)}) \;&\le\; r^\gamma b_d^{\gamma/d}\Bigg(\Lambda(\mathcal{L}_{B_r(x)})^{\gamma/\beta}\sup\Big\{y^{-\gamma/\beta}(y-1+e^{-y})\colon \\ &\qquad 0 \le y \le \Lambda(\mathcal{L}_{B_r(x)})\Big\} \\ &\qquad + \sup\left\{y^{\gamma-\beta}\left|\int_{\mathcal{L}_{B_r(x)}}\phi(u)\left(L(y^{1/\eta}u)-1\right)du\right|\colon y \ge 1\right\}\Bigg) \\ &= \; b_d^{\gamma/d}\Bigg(C\left(\frac{\gamma}{\beta};\, r^{-\beta}\Lambda(\mathcal{L}_{B_1(x/r)})\right)\Lambda(\mathcal{L}_{B_1(x/r)})^{\gamma/\beta} \\ &\qquad + \kappa(\mathcal{L}_{B_1(x/r)}; r)\Bigg),\end{aligned}$$

where

$$C(x; z) = \sup\{y^{-x}(y-1+e^{-y})\colon 0 \le y \le z\}$$

and

$$\kappa(\mathcal{L}_K; r) = \sup\left\{y^{\gamma-\beta}\left|\int_{\mathcal{L}_K}\phi(u)\left(L(y^{1/\eta}u)-1\right)du\right|\colon y \ge r\right\}.$$

Finally, estimate the distance $\mathfrak{r}_\Phi(A_1, X; \mathcal{M}(a, R))$ by

$$\begin{aligned}&\mathfrak{r}_\Phi(A_1, X; \mathcal{M}(a, R)) = \\ &\quad = \sup\{\mathfrak{r}_\Phi(A_1, X; \mathcal{M}'_K)\colon K \in \mathcal{M}_0(a, R)\} \\ &\quad \le b_d^{\gamma/d}\sup\Bigg\{\Bigg[C\left(\frac{\gamma}{\beta}; r^{-\beta}\Lambda(\mathcal{L}_{B_1(x/r)})\right)\Lambda(\mathcal{L}_{B_1(x/r)})^{\gamma/\beta} \\ &\qquad\qquad + \kappa(\mathcal{L}_{B_1(x/r)}; r)\Bigg)\colon 0 \le r \le \frac{R-a}{2}, 1 + \frac{a}{r} \le \|v\|\Bigg\}.\end{aligned}$$

Unfortunately, this estimate tends to infinity as $R \to \infty$ or $a \downarrow 0$.

In particular, let $M(\xi) = \{\xi\}$ be an isotropic random set (singleton), and let $L(u)$ be equal to 1 outside $B_a(0)$. Then

$$\mathfrak{r}_\Phi(A_1, X; \mathcal{M}(a, R)) \leq b_d^{\gamma/d} C(\gamma/\beta; s_0) \Lambda(B_1(qe))^{\gamma/\beta},$$

where e is a certain unit vector,

$$q = 1 + \frac{2a}{R - a}$$

and

$$s_0 = \sup \left\{ r^{-\beta} \Lambda \left(B_1(0) + (1 + \frac{a}{r})e \right) : 0 \leq r \leq \frac{R - a}{2} \right\}.$$

Chapter 8

Applications of Limit Theorems

8.1 Simulation of Stable Random Sets.

Many difficulties in the theory of random sets and in statistics of random sets are caused by the shortage of models or distributions of random sets. A good model of random sets should be easy to simulate and have a wide spectrum of possible shapes of realizations. It is desirable to have an explicit formula for its distribution (e.g., for its capacity functional), which could be computed without serious difficulties.

In fact, until now only one such a model of random sets is known. It is the well-known *Boolean model* of random sets, see Matheron (1975), Serra (1982). The Boolean model is applicable in many branches of natural science, see Stoyan et al. (1987) and references therein.

The Boolean model A is defined by

$$A = \bigcup_{x_i \in \Pi_\Lambda} (x_i + A_0^i), \tag{1.1}$$

where Π_Λ is the Poisson point process in $\mathbb{R}^d$ with the intensity measure Λ, the A_0^i, $i \geq 1$, form a sequence of iid random sets. These sets are assumed to be copies of a certain random closed set A_0 called the *"typical" grain* of the Boolean model A.

The simulation of a Boolean model within a set W falls into the following stages. First, the number of points is determined by simulating a Poisson random variable N with parameter $\Lambda(W_1)$. The set W_1 is the enlarged window defined as $W_1 = W \oplus B_R(0)$ for sufficiently large R. (If A_0 is bounded a.s., then it is advisable to put $R \geq \|A_0\|$.) Then N independent random points in W_1 with the distribution $\Lambda(dx)/\Lambda(W_1)$ and N iid copies of A_0 are to be simulated. Finally, A is constructed by (1.1). The "typical" grain A_0 is supposed to have a simple distribution. In many applications the grain is chosen to be a disc, polygon or ellipse of random size and/or orientation etc. Even Boolean models with deterministic grains are sometimes considered.

The capacity functional of the Boolean model A is equal to

$$T(K) = 1 - \exp\left\{-\mathbf{E}\Lambda(A_0 \oplus \check{K})\right\}, \tag{1.2}$$

see also Example 3.2.2.

A generalization of Robbins formula for expected volumes of random sets (see Robbins, 1944, Matheron 1975) yields

$$T(K) = 1 - \exp\left\{-\int_{\mathbb{R}^d} T_{A_0}(K + x)\Lambda(dx)\right\}, \tag{1.3}$$

where $T_{A_0}(K)$ is the capacity functional of A_0. Note that (1.3) may be obtained very simple from the Fubini theorem.

Although the Boolean model is a convenient model and is applied to many practical problems, new models of random sets are most appreciated to be derived. There is a special need of models of random compact sets, since they can be used as models for the typical grain A_0 of a certain Boolean model. In this connection U-stable and C-stable random sets may serve as relevant models of random closed sets. It follows from limit theorems for unions that U-stable sets can be simulated with given accuracy by means of an appropriate sequence of iid random sets. Besides, capacity functionals of stable random sets are expressed by explicit formulae, see Chapter 4. Below we mention several examples confirming that U-stable sets may be of various shapes.

It should be noted that sometimes the Boolean model is U-stable. For example, this is true in case Λ is a homogeneous measure and A_0 is a single-point random set (random point). Then A coincides with the Poisson point process Π_Λ, see Examples 3.2.1 and 3.2.2.

In general, U-stable sets can be simulated by means of a certain sequence $A_1, A_2, \dots$ of iid random closed sets and scale transformation of their union, that is we put

$$X_n = a_n^{-1}(A_1 \cup \dots \cup A_n) \tag{1.4}$$

for suitable constants a_n, $n \geq 1$. In practice, the random sets $A_1, A_2, \dots$ (summands) are supposed to have a simple distribution. Since, in fact, we can simulate random variables and random vectors only, the scheme of Theorem 4.4.5 is applicable. Then A_1 is supposed to be equal to $M(\xi)$, where ξ is a random vector in $\mathbb{R}^m$, and $M: \mathbb{R}^m \longrightarrow \mathcal{K}$ is a multivalued homogeneous function with compact values in $\mathbb{R}^d$. Below we consider random closed sets in $\mathbb{R}^2$ only (i.e. $d = 2$).

In the simplest case M is a single-point-valued function, i.e. $M(\xi) = \{\xi\}$. Here ξ is assumed to have the probability density

$$f(x) = \frac{C(m,\alpha)}{1 + \|x\|^{m-\alpha}}, \quad d = m = 2, \tag{1.5}$$

for a suitable constant $C(m,\alpha)$. In fact,

$$C(m,\alpha) = \Gamma\left(\frac{m}{2}\right)(m-\alpha)\pi^{-m/(m-\alpha)} \sin\left(\frac{m\pi}{m-\alpha}\right), \tag{1.6}$$

since the integral of $f(x)$ is to be equal to 1. Evidently, the density f is spherically symmetric. It is also regularly varying with $\operatorname{ind} f = \alpha - m$.

By (4.4.13), the norming constants are to be equal to

$$a_n = (C(2,\alpha)n)^{-1/\alpha}.$$

Furthermore, the limiting random set X has the capacity functional $\tilde{T}$ given by

$$\tilde{T}(K) = 1 - \exp\left\{-\int_K \|u\|^{\alpha-2} du\right\}. \tag{1.7}$$

Figure 1 presents simulation results of the set X_n from (1.4) for $\alpha = -2$ and several values of n ($n = 100, 1000, 10000$).

Figure 1

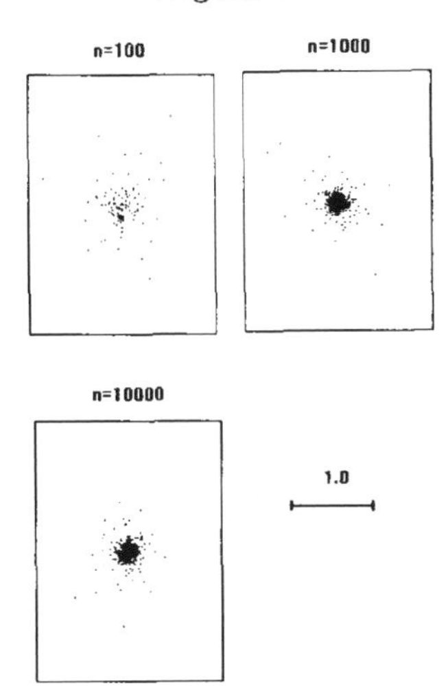

It should be noted that all sets on Figure 1 consist of points only. They are countable and non-fractal. Since f is rotation invariant, random sets on Figure 1 are isotropic.

An example of a non-isotropic U-stable set can be constructed similarly by taking

$$f(x) = \frac{C}{1 + \phi(x\|x\|^{-1})\|x\|^{m-\alpha}}.$$

Such a random set can be simulated either directly or by means of the transformation $x \longrightarrow x\phi(x/\|x\|)$ applied to the isotropic random sets from Figure 1.

Assume that $m = 6$ and $M(\xi) = M(\xi_1, \ldots, \xi_6)$ is a random triangle having the vertices (ξ_1, ξ_2), (ξ_3, ξ_4) and (ξ_5, ξ_6). In this case M is homogeneous of degree $\eta = 1$. Put $a_n = n^{-1/\alpha}$, where

$$f(u) = \frac{C(m, \alpha)}{1 + \|u\|^{m-\alpha}}, u \in \mathbb{R}^m, m = 6. \tag{1.8}$$

is the density of $\xi = (\xi_1, \ldots, \xi_6)$. Then the capacity functional of the limiting random set X in the union scheme is equal to

$$\tilde{T}(K) = 1 - \exp\left\{-C(6, \alpha) \int_{\mathcal{L}_K} \|u\|^{\alpha-6} du\right\}, 0 \notin K. \tag{1.9}$$

Simulation results are presented in Figure 2 for the parameters $\alpha = -1$ and $\alpha = -3$. The number n is equal to 200.

Furthermore, we can consider the Boolean model whose typical grain A_0 coincides (in distribution) with a certain U-stable random set X, say presented on Figure 2.

Figure 2

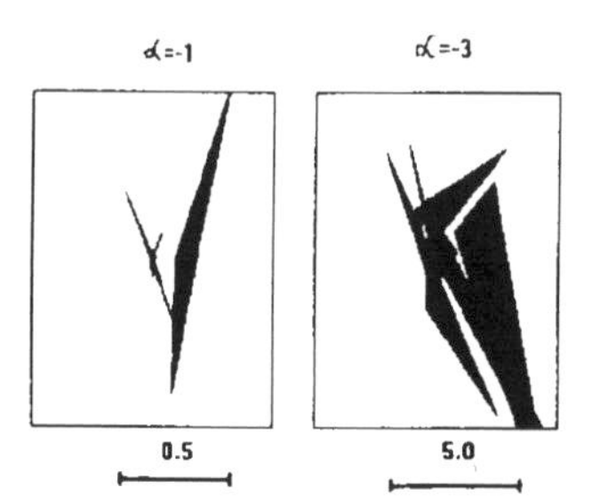

A sample of such a Boolean model simulated by the stationary Poisson point process with intensity $\lambda = 0.0055$ is given in Figure 3 within the unit 60×55 for $\alpha = -3$. Here each copy of X is produced by an appropriate scale transformation of the union of $n = 100$ triangles.

Note that the Boolean model A has the capacity functional

$$T(K) = 1 - \exp\left\{-\lambda \int_{\mathbb{R}^d} \left(1 - \exp\left\{-C(6,\alpha) \int_{\mathcal{L}_{K+x}} \phi(u)du\right\}\right) dx\right\}, \tag{1.10}$$

where the function ϕ is given by

$$\phi(u) = \|u\|^{\alpha-6}.$$

Figure 3

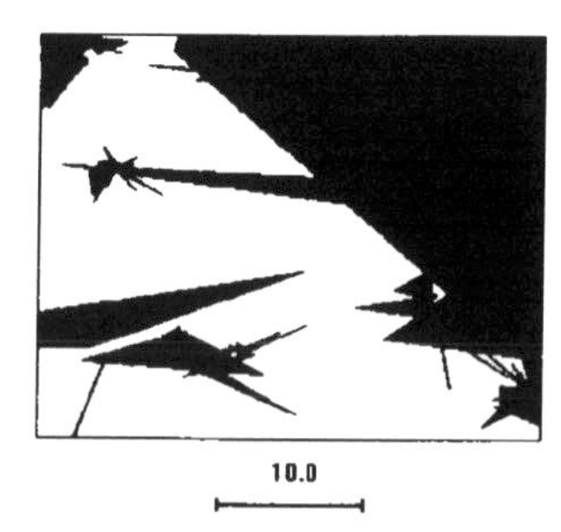

Let $M(\xi) = M(\xi_1, \xi_2, \xi_3)$ be the ball of radius ξ_3 centered at (ξ_1, ξ_2), and let ξ have the density

$$f(u) = 2\frac{C(m,\alpha)}{1 + \|u\|^{m-\alpha}}, \quad m = 3, d = 2, \tag{1.11}$$

for $u = (u_1, u_2, u_3)$, $u_3 \geq 0$. In this case the radius and the center of the ball are dependent and their common distribution is spherically symmetric.

Figure 4 shows two samples of the set

$$X_n = a_n^{-1}(A_1 \cup \cdots \cup A_n),$$

where $A_1, \ldots, A_n$ are iid copies of $M(\xi)$, $a_n = (2C(m,\alpha)n)^{-1/\alpha}$ and $\alpha = -3$. The number of "summands" n is equal to 100 and 1000 respectively.

Figure 4

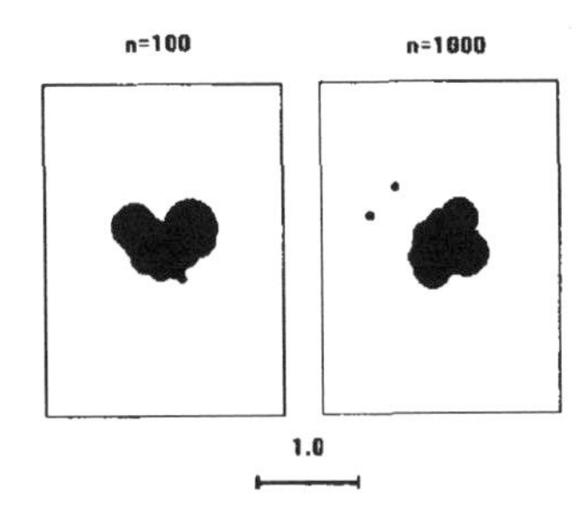

It should be noted that the accuracy of simulation can be estimated by means of the probability metrics method, see Chapter 7. This technique may be used to estimate the uniform or Levy distances between the capacity functional of X_n and the limiting random set X, see Section 7.4.

Convex-stable sets are simulated similarly as convex hulls of union-stable random closed sets. The corresponding pictures can be obtained by taking convex hulls of sets shown on Figures 1,2 and 4. It was proven in Davis, Mulrow and Resnick (1987) that the convex hull for the setting of Figure 1 has almostsurely a finite number of vertices.

Our models of union-stable and convex-stable random sets are likely useful for modeling and description of objects in many practical examples.

EXAMPLE **1.1** Assume that a random vector ξ represents the location of a star in a star-cluster. Then the star-cluster itself is the union of stars (or their convex hull). After scale transformation we obtain a U-stable (respectively C-stable) random closed set. Such sets are depicted in Figure 1 for ξ having the density (1.5) and several values of n.

EXAMPLE **1.2** Suppose a certain disease has appeared in circular elementary regions $A_1, A_2, \ldots$ on a map. The origin stands for the primary center of the disease propagation. Then scaled unions of these elementary regions tend to a U-stable random set. Of course, their scaled convex hull is C-stable. It might be appropriate to assume that the conex hull describes the region to be placed in quarantine. We can apply Theorem 4.4.8 with $m = 3$, $d = 2$ and $M = B_1(0)$. Hence the weak limit for unions with suitable norming factors a_n is a U-stable set X with the capacity functional given by

$$\tilde{T}(K) = 1 - \exp\left\{-\int_0^\infty dy \int_{K^y} \phi(u;y)du\right\}, \tag{1.12}$$

where $K^y = K \oplus B_y(0) \subset \mathbb{R}^2$ and $f = \phi L$ is the representation (1.6.8) of the regularly varying density of the random vector (ξ_1, ξ_2, ξ_3). Here an elementary region is defined as $A_1 = (\xi_1, \xi_2) + \xi_3 M$. For f defined in (1.11) we get

$$\phi(u; y) = \phi(u_1, u_2, y) = \left(\|u\|^2 + y^2\right)^{(\alpha-d)/2},$$

and $a_n = (2nC(m, \alpha))^{-1/\alpha}$, see (1.6). The corresponding examples of simulated sets are shown in Figure 4. The scaled convex hull $a_n^{-1} Z_n$ of these elementary regions $A_1, A_2, \ldots$ converges weakly to the random C-stable set Z with the inclusion functional

$$\tilde{\imath}(K) = \exp\left\{-\int_0^\infty dy \int_{(F^{-y})^c} \phi(u; y) dy\right\}, \tag{1.13}$$

where $F^{-y} = F \ominus B_y(0)$. The support function of the Aumann expectation of Z is equal to

$$s_{\mathbf{E}Z}(v) = \Gamma(1-\gamma)\left[\gamma \int_0^\infty dy \int_{S_v^+} \phi(w; y)\left((w \cdot v) + y\right)^{1/\gamma} dw\right]^\gamma, \quad v \in \mathbb{S}^{d-1}, \tag{1.14}$$

where $\gamma = -1/\alpha$, $S_v^+ = \{u \in \mathbb{S}^{d-1} : (u \cdot v) \geq 0\}$ and $d = 2$.

Similarly, original sets $A_1, A_2, \ldots$ (summands) can be used to describe polluted regions (e.g., oil spots in the sea). Then their scaled union is union-stable, so that U-stable random sets may be of use for the simulation pollution propagation. In this case the origin stands for the primary source of pollution.

EXAMPLE **1.3** Let $A_1 = [0, \xi]$ be the segment with one end-point in the origin and the other one in the random point ξ in $\mathbb{R}^2$. This set may represent the direction and the strength of the wind at a fixed place. Then $X_n = a_n^{-1}(A_1 \cup \cdots A_n)$ represents the extremal rose of directions of the wind (compare with the Minkowski sum of $A_1, A_2, \ldots, A_n$, which is the usual rose of directions).

In order to find the capacity functional of the corresponding weak limit of X_n we can apply Theorem 4.4.5 to the multivalued function $M(u) = [0, u]$. Then

$$\begin{aligned} \mathcal{L}_K &= \{u : M(u) \cap K \neq \emptyset\} \\ &= \{cx : x \in K, c \geq 1\} = \hat{K}. \end{aligned}$$

If ξ admits a regularly varying density $f = \phi L$, then X_n converges weakly to the random closed set X with the capacity functional

$$\tilde{T}(K) = 1 - \exp\left\{-\int_{\hat{K}} \phi(u) du\right\}, K \in \mathcal{K}.$$

A sample of such a set is obtained from the sets given in Figure 1 after joining all points with the origin. The expected convex hull of scaled unions describes the rose of extremal directions too. Its support function is given by (4.4.8). For ξ having the density (1.5) we get

$$s_{\mathbf{E}Z}(v) = \Gamma(1 + 1/\alpha)\left[-\frac{1}{\alpha}\int_{S_v^+} \|u\|^{\alpha-d}(w, v)^{-\alpha} du\right]^{-1/\alpha}, v \in \mathbb{S}^{d-1}, d = 2.$$

Stable random sets with positive indices come from the scheme of Theorem 4.4.11. Consider the following example, which may be of use, say for the study of corrosion propagation.

EXAMPLE 1.4 Let $\mathbb{C}$ be a cone in $\mathbb{R}^d$, and let us suppose that random points $\xi_1, \xi_2, \dots$ describe defects in this cone. Then the random convex set $Z_n = \mathrm{conv}\{\xi_1, \dots, \xi_n\}$ may be interpreted as the destroyed region. Its limit distribution is defined by (4.4.23). For example, take $d = 2$ and $\mathbb{C} = \{(u_1, u_2) : u_1, u_2 \geq 0\}$. Then

$$\mathfrak{r}(F) = \exp\left\{-\frac{1}{\alpha}\int_0^\pi \tilde{\phi}(\beta)\tilde{\rho}(\beta)d\beta\right\},$$

where

$$\begin{aligned} \tilde{\phi}(\beta) &= \phi(\cos\beta, \sin\beta), \\ \tilde{\rho}(\beta) &= \inf\{x > 0 : (x\cos\beta, x\sin\beta) \in F\}. \end{aligned}$$

Thus, the expectation of the limiting random set Z is given by means of the corresponding mean support function

$$\mathbf{E}s_Z(-v) = -\Gamma\left(1 + \frac{1}{\alpha}\right)\left(-\frac{1}{\alpha}\int_0^{\pi/2} \tilde{\phi}(\beta)\cos(\beta - \kappa)d\beta\right)^{-1/\alpha}$$

for $v = (\cos\kappa, \sin\kappa)$ and $0 \leq \kappa \leq \pi/2$. For other v the left-hand side is infinite. If $\tilde{\phi}(\beta) = c$, $0 \leq \beta \leq \pi/2$ (the angle distribution of ξ_1 is uniform within $\mathbb{C}$), then

$$\mathbf{E}s_Z(-v) = -\Gamma(1 + 1/\alpha)\left(-c\alpha^{-1}(\cos\kappa + \sin\kappa)\right)^{-1/\alpha}.$$

8.2 Estimation of Tail Probabilities for Volumes of Random Samples.

The limit behavior of functionals of convex hulls of random samples is an important subject of stochastic geometry (see Schneider, 1988). Limit theorems for convex hulls of random sets imply weak convergence of any continuous and bounded functional of normalized convex hulls to its value on the limiting C-stable random closed set. The simplest functional is the volume of a convex set. The unboundedness of this functional is overcome by considering truncated random sets, i.e. intersections of scaled convex hulls with a certain non-random ball. Below we deal with volumes of random samples and their convex hulls.

Following Vitale (1988), consider the similar problem for *Gaussian* random samples. It was proven in Vitale (1988) that

$$\mathbf{E}\mu^{1/d}(X) \leq \mu^{1/d}(\mathbf{E}X) \tag{2.1}$$

for any random set X with $\mathbf{E}\|X\| < \infty$. Here $\|X\| = \sup\{\|u\| : u \in X\}$, $\mathbf{E}X$ is the Aumann expectation of X (see Chapter 2), and μ is the Lebesgue measure in $\mathbb{R}^d$.

Suppose that $\xi_1, \dots, \xi_n$ are iid random points in $\mathbb{R}^d$ each having standard Gaussian distribution and X_n is their convex hull. Then the expectation $\mathbf{E}X_n$ is equal to $a_n B$, where B is the unit ball and a_n is the expected value of the maximum of n standard Gaussian variables in $\mathbb{R}^1$. Then, for each $x > 0$,

$$\begin{aligned} \mathbf{P}\{\mu(X) \geq x\} &= \mathbf{P}\{\mu^{1/d}(X_n) \geq x^{1/d}\} \\ &\leq x^{-1/d}\mathbf{E}\mu^{1/d}(X_n) \\ &\leq a_n(b_d/x)^{1/d}, \end{aligned} \tag{2.2}$$

where $b_d = \mu(B)$ is the volume of the unit ball.

We derive the similar inequality for general random samples, provided the conditions of limit theorems for convex hulls are satisfied. Let $A_1, \dots, A_n$ be iid copies of a certain random closed set A. Suppose that their convex hull

$$a_n^{-1}Z_n = a_n^{-1}\mathrm{conv}(A_1 \cup \dots \cup A_n)$$

converges weakly to a certain C-stable random set Z with parameter α. The reader is referred to the notations introduced in Chapter 4.

Theorem 2.1 *Let* $\mathbf{E}Z$ *be compact. Then for any* $R, \varepsilon > 0$ *and sufficiently large* n

$$\mathbf{P}\{\mu(\mathrm{conv}(A_1 \cup \dots \cup A_n) \cap a_n B_R(0)) \geq x\} \leq a_n \left[\frac{\mu(\mathbf{E}Z) + \varepsilon}{x}\right]^{1/d}, \quad x > 0. \tag{2.3}$$

PROOF. It follows from (2.1) that for any $R > 0$

$$\mathbf{E}\mu^{1/d}\left(a_n^{-1}Z_n \cap B_R(0)\right) \leq \mu^{1/d}\left(\mathbf{E}\left[a_n^{-1}Z_n \cap B_R(0)\right]\right). \tag{2.4}$$

The functional $F \longrightarrow \mu(F \cap B_R(0))$ is bounded and continuous for convex F in the Hausdorff metric. Theorem 1.4.6 yields

$$\mu\left(\mathbf{E}\left[a_n^{-1}Z_n \cap B_R(0)\right]\right) \to \mu\left(\mathbf{E}[Z \cap B_R(0)]\right) \text{ a.s. as } n \to \infty.$$

Then, by (2.4) and Markov's inequality, we get

$$\begin{aligned} &\mathbf{P}\{\mu(\mathrm{conv}(A_1 \cup \dots \cup A_n) \cap a_n B_R(0)) \geq x\} = \\ &= \mathbf{P}\left\{\mu\left(a_n^{-1}Z_n \cap B_R(0)\right)^{1/d} \geq a_n^{-1}x^{1/d}\right\} \\ &\leq \frac{\mathbf{E}\mu^{1/d}\left(a_n^{-1}Z_n \cap B_R(0)\right)}{a_n^{-1}x^{1/d}} \\ &\leq a_n\left[\frac{\mu\left(\mathbf{E}(a_n^{-1}Z_n \cap B_R(0)\right)}{x}\right]^{1/d} \\ &\leq a_n\left[\frac{\mu(\mathbf{E}Z) + \varepsilon}{x}\right]^{1/d}. \quad \square \end{aligned}$$

It follows from (2.3) that

$$\mathbf{P}\left\{\mu\left(a_n^{-1}Z_n \cap B_R(0)\right) \geq x\right\} \leq \left[\frac{\mu(\mathbf{E}Z) + \varepsilon}{x}\right]^{1/d}, \quad x > 0, \tag{2.5}$$

for all sufficiently large n.

The expectation $\mathbf{E}Z$ was computed in Section 4.4 for several special examples of random sets $A_1, A_2, \ldots$. It is determined by the corresponding support function $s_{\mathbf{E}Z}$. If $d = 2$, then the volume $\mu(\mathbf{E}Z)$ is evaluated by

$$\mu(\mathbf{E}Z) = 1/2 \int_0^{2\pi} \left(p(\beta)^2 - p'(\beta)^2\right) d\beta, \tag{2.6}$$

see Santaló (1976). Here $p(\beta) = s_{\mathbf{E}Z}(\cos\beta, \sin\beta)$, $0 \le \beta \le 2\pi$, is the support function of $\mathbf{E}Z$. For general d

$$\mu(\mathbf{E}Z) \le \frac{1}{d2^{d-1}} \int_{\mathbb{S}^{d-1}} s_{\mathbf{E}Z}(u)du, \tag{2.7}$$

see Uryson (1924).

Now consider convex hulls of random points. In this case $A_1 = \{\xi\}$ is a single-point-random set. Then (2.3) and (2.5) estimate the tail probabilities of distribution of $\mu(\text{conv}\{\xi_1, \ldots, \xi_n\})$.

Suppose that the conditions of Theorems 4.4.1 and 4.4.3 are valid. Then the normalized random sample

$$a_n^{-1}\text{conv}\{\xi_1, \ldots, \xi_n\}$$

converges weakly to the C-stable random set Z with the inclusion functional

$$\mathfrak{i}(F) = \exp\left\{-\int_{F^c} \phi(u)du\right\}, \quad F \in \mathcal{C}, \tag{2.8}$$

The expectation of Z is determined by

$$s_{\mathbf{E}Z}(v) = \Gamma(1-\gamma)\left[\gamma \int_{S_v^+} \phi(u)(w \cdot v)^{1/\gamma}du\right]^{\gamma}, \quad v \in \mathbb{S}^{d-1}, \tag{2.9}$$

where $\gamma = -1/\alpha$, $\alpha < -1$, $S_v^+ = \{u \in \mathbb{S}^{d-1} : (u \cdot v) \ge 0\}$ and $(\alpha - d)$ is the index of the regularly varying density f of ξ_1. Suppose that $f(tu) \sim \phi(tu)$ as $t \to \infty$. Then $a_n = n^{\gamma}$. Thus

$$\lim_{n\to\infty} \mathbf{P}\left\{\mu\left(n^{-\gamma}Z_n \cap B_R(0)\right) \ge x\right\} \le \left[\frac{\mu(\mathbf{E}Z) + \varepsilon}{x}\right]^{1/d}, x > 0. \tag{2.10}$$

Let ϕ be circular symmetric, i.e. $\phi(u) = C$ for all $u \in \mathbb{S}^{d-1}$. Then $\mathbf{E}Z = B_r(0)$, where

$$r = \Gamma(1-\gamma)\left[\gamma C \int_{S_v^+} (u \cdot v)^{1/\gamma}du\right]^{\gamma}. \tag{2.11}$$

It follows from (2.10) that

$$\lim_{n\to\infty} \mathbf{P}\left\{\mu\left(n^{-\gamma}Z_n \cap B_R(0)\right) \ge x\right\} \le b_d^{1/d} r x^{-1/d}, \ x > 0. \tag{2.12}$$

For planar samples ($d = 2$) we get

$$\lim_{n\to\infty} \mathbf{P}\left\{\mu\left(n^{-\gamma}Z_n \cap B_R(0)\right) \ge x\right\} \le \left(\frac{\pi}{x}\right)^{1/2} r$$

and

$$r = \Gamma(1-\gamma)\left[\gamma C \int_{-\pi/2}^{\pi/2} (\cos\beta)^{1/\gamma} d\beta\right]^{\gamma}.$$

For example, if $\alpha = -2$, then $r = \pi C^{1/2}/2$.

The above estimators for tail probabilities for volumes of random samples are useful in constructing statistical tests for C-stable random sets.

It was explained in Example 1.1 that C-stable sets may describe distributions of normalized convex hulls of star-clusters, provided the location of a star has regularly varying density f. Now we can check also whether or not a collection of stars forms *one* or *several* clusters, provided the density f is known. It can be done as follows. For simplicity suppose that $f(xu) \sim \phi(xu)$ as $x \to \infty$ and $u \in \mathbb{R}^d \setminus \{0\}$. Here ϕ is a homogeneous function.

First, evaluate the expectation of the limiting C-stable random set Z by Theorem 4.4.3 and its volume $\mu(\mathbf{E}Z)$ (e.g., by (2.6) or (2.7)). Then compute

$$\zeta = n^{\gamma}\left(\frac{\mu(\mathbf{E}Z)}{V_n}\right)^{1/d}, \tag{2.13}$$

where V_n is the volume of the convex hull Z_n of the collection of n random locations of stars (or random points). If ζ is smal, then the collection of stars unlikely forms a unique cluster.

The values of ζ were computed for random samples in $\mathbb{R}^2$ having the density f given by (1.5) with $\alpha = -2$. In this case $\zeta = (\mu(\mathbf{E}Z)/nV_n)^{1/2}$, and $\mathbf{E}Z$ is the ball $B_r(0)$ with $r = 2^{-1/4}$. It has been done for 100 independent samples of sets $a_n^{-1}\{\xi_1, \ldots, \xi_n\}$ each having n points. The empirical counterparts $\hat{x}_q(n)$ of quantiles x_q of ζ defined by $\mathbf{P}\{\zeta \le x_q\} = q$ are presented in Table 1.

Table 1: Empirical quantiles $\hat{x}_q(n)$ and expected values of ζ evaluated by 100 independent samples, each having n points.

n \ q	0.01	0.025	0.05	0.1	0.2	0.3	0.9	$\mathbf{E}Z$
100	0.88	0.94	1.02	1.05	1.20	1.33	2.06	1.55
1000	0.80	0.85	0.91	0.98	1.22	1.28	2.12	1.52
10000	0.98	1.05	1.16	1.28	1.38	1.51	2.29	1.71

Now consider unions of random closed sets with positive volumes. In this case the expected value of the volume of the limiting random set are to be computed.

Let $X_n = A_1 \cup \cdots \cup A_n$, and let X be the limiting random set of $a_n^{-1}X_n$ for suitable constants a_n, $n \ge 1$. Suppose that the conditions of Theorem 4.4.5 are valid. Robbins formula (see, e.g., Matheron, 1975) yields

$$\mathbf{E}\mu(X) = \int_{\mathbb{R}^d} \mathbf{P}\{x \in X\}\, dx.$$

From (4.4.14) we get

$$\begin{aligned} \mathbf{E}\mu(X) &= \int_{\mathbb{R}^d} \tilde{T}(\{x\})dx \\ &= \int_{\mathbb{R}^d} \left(1 - \exp\left\{-\int_{\mathcal{L}_x} \phi(u)du\right\}\right) dx, \end{aligned}$$

where $\mathcal{L}_x = \{u \in \mathbb{R}^m : x \in M(u)\}$. Recall that $M: \mathbb{R}^m \longrightarrow \mathcal{K}$ is a multivalued function whose values are closed compacts in $\mathbb{R}^d$. Suppose that $\eta d/\alpha > -1$, i.e. $\alpha < -\eta d$. If $x = ye$ for some $y \geq 0$ and $e \in \mathbb{S}^{d-1}$, then $\mathcal{L}_x = y^{1/\eta}\mathcal{L}_e$. Hence

$$\begin{aligned} \mathbf{E}\mu(X) &= \int_{\mathbb{S}^{d-1}} \mu_{d-1}(de) \int_0^\infty \left(1 - \exp\left\{-y^{\alpha/\eta} \int_{\mathcal{L}_e} \phi(u)du\right\}\right) y^{d-1}dy \\ &= \Gamma\left(\frac{\eta d}{\alpha} + 1\right) \int_{\mathbb{S}^{d-1}} \left[\int_{\mathcal{L}_e} \phi(u)du\right]^{-\eta d/\alpha} \mu_{d-1}(de), \end{aligned} \tag{2.14}$$

where μ_{d-1} is the uniform measure on $\mathbb{S}^{d-1}$.

If ξ from Theorem 4.4.5 have a circular symmetric distribution, then

$$\mathbf{E}\mu(X) = \frac{2\pi^{d/2}\Gamma(\eta d/\alpha + 1)}{d\Gamma(d/2)} \left[\int_{\mathcal{L}_e} \phi(u)du\right]^{-\eta d/\alpha}.$$

Note that $\mathbf{E}\mu(X)$ does not depend on e belonging to $\mathbb{S}^{d-1}$.

For $v \in \mathbb{R}^m$ denote

$$\begin{aligned} q_1(e,v) &= \inf\{y \geq 0:\ ye \in M(v)\}, \\ q_2(e,v) &= \sup\{y \geq 0:\ ye \in M(v)\}. \end{aligned}$$

In case the set in question is empty we put $q_1(e,v) = 0$ and $q_2(e,v) = +\infty$. Then

$$\int_{\mathcal{L}_e} \phi(u)du = \int_{\mathbb{S}^{m-1}} \phi(v)\left[q_2(e,v)^{-\alpha/\eta} - q_1(e,v)^{-\alpha/\eta}\right] \mu_{m-1}(dv).$$

Finally, for any positive b, R, ε and sufficiently large n, we get

$$\begin{aligned} &\mathbf{P}\{\mu((A_1 \cup \cdots \cup A_n) \cap a_n B_R(0)) \geq b\} = \\ &\quad = \mathbf{P}\left\{\mu\left(a_n^{-1}X_n \cap B_R(0)\right)^{1/d} \geq a_n^{-d}b\right\} \\ &\quad \leq \frac{\mathbf{E}\mu\left(a_n^{-1}X_n \cap B_R(0)\right)}{a_n^{-d}b} \\ &\quad \leq \frac{\mu(\mathbf{E}Z) + \varepsilon}{a_n^{-d}b}. \end{aligned}$$

8.3 Convergence of Random Sets Generated by Graphs of Random Functions.

In this section we apply limit theorems for random sets to unions of hypographs of random upper semi-continuous functions. It should be noted that limit theorems for unions of hypographs imply similar limit theorems for maximums of random functions. Related results for extremes of random processes can be found in Norberg (1986b,1987).

A real-valued function $g: E \longrightarrow \mathbb{R}$ on a locally compact Hausdorff separable linear space E is said to be *upper semi-continuous* if $\limsup_{x\to a} g(x) = g(a)$ for all $a \in E$. (In the following E is considered to be the Euclidean space $\mathbb{R}^d$.) It can be shown that g is upper semi-continuous if and only if its *hypograph*

$$\text{hypo}g = \{(x,t)\colon t \le g(x), x \in E, t \in \mathbb{R}\} \tag{3.1}$$

is closed in $E \times \mathbb{R}$ furnished with the product-topology.

If ξ is a random function with almost surely upper semi-continuous realizations, then hypoξ is a random closed subset of $E \times \mathbb{R}$. Thus, basic concepts of random sets theory can be reformulated for random functions. Following Molchanov (1993a), we say that ζ_n *hypo-converges* to ζ if hypoζ_n converges weakly to hypoζ as random closed sets, cf. Salinetti and Wets (1986), Attouch and Wets (1990), where this idea was introduced and discussed for epigraphs of random lower semi-continuous functions.

Let $\xi_1, \xi_2, \ldots$ be iid copies of a certain upper semi-continuous random function ξ. In the hypographical language pointwise maxima of functions turn into unions of their hypographs. Namely,

$$\text{hypo}\,(\max(\xi_1,\ldots,\xi_n)) = \bigcup_{i=1}^{n} \text{hypo}\xi_i. \tag{3.2}$$

To proceed further we normalize unions of hypographs or maxima of random processes. In previous chapters we use only scale transformation. However, hypographs of random functions are subsets of the Cartesian product $E \times \mathbb{R}$. Hence scale factors for parameters and values are naturally allowed to be different from each other. Define

$$\zeta_n(u) = c_n^{-1} \max\left(\xi_1(b_n u), \ldots, \xi_n(b_n u)\right) \tag{3.3}$$

for certain positive constants c_n and b_n. Thus

$$Z_n = \text{hypo}\zeta_n(x) = a_n^{-1} \circ \bigcup_{i=1}^{n} \text{hypo}\xi_i, \tag{3.4}$$

where $a_n = (b_n, c_n) \in \mathbb{R}_+^2 = \mathbb{C}$ and

$$a_n^{-1} \circ F = \left\{(b_n^{-1}x_1, c_n^{-1}x_2)\colon (x_1,x_2) \in F\right\}, \quad F \subset E \times \mathbb{R}, \tag{3.5}$$

cf. Section 4.5. Thus, ζ_n hypo-converges to a certain random function ζ if the random closed set A = hypoξ satisfies the conditions of Theorem 4.5.1. It means that the function

$$\tau_K(x) = T_A(x \circ K) = \mathbf{P}\left\{A \cap x \circ K \neq \emptyset\right\}, x = (x_1, x_2)$$

is regularly varying on $\mathbb{C}$ for each K such that $\liminf_{x\to\infty} T(x \circ K) = 0$.

Besides,

$$q_n(K) = \sup\{t \geq 0: \tau_K(ta_n) \geq 1/n\} \tag{3.6}$$

should have a limit (maybe infinite).

General results on the weak convergence of random closed sets are presented in Norberg (1984), Salinetti and Wets (1986) (see also Section 1.4 for a brief discussion). It is known that pointwise convergence of capacity functionals on the class $\mathcal{K}_{up}$ of all finite unions of parallelepipeds implies the weak convergence of the corresponding random closed sets. The class $\mathcal{K}_{up}$ consists of compacts of the form

$$K = \bigcup_{i=1}^{n} K_i \times [c_i, d_i], \tag{3.7}$$

where K_i, $1 \leq i \leq m$, are compact subsets of E. We can safely suppose that $K_1, \ldots, K_m$ are convex and their intersections are pair-wise disjoint. Since hypoξ consists of entire vertical half-lines, we may put $d_i = \infty$.

Now we check conditions of Theorem 4.5.1 for such a compact K and a special example of random function ξ. Put

$$\xi(u) = \eta_2 g(u - \eta_1), \ u \in \mathbb{R}^d, \tag{3.8}$$

where $g: \mathbb{R}^d \longrightarrow [0, \infty)$ is a bounded continuous non-random function, η_1 is a random vector in $\mathbb{R}^d$, and η_2 is a positive random variable independent with η_1. Suppose that η_1 and η_2 have regularly varying densities f_1 and f_2 admitting representations $f_i = \phi_i L_i$, $i = 1, 2$. Moreover, let ind$f_1 = \alpha_1 - d$ and ind$f_2 = \alpha_2 - 1$ for a certain negative constants α_1 and α_2.

We assume also that the origin is the maximum point of g, that is

$$g(0) = \sup\{g(u): \ u \in \mathbb{R}^d\} = g_0,$$

and

$$\tilde{g}(r) = \sup_{u \notin B_r(0)} g(u) = o(r^\gamma) \ \text{ as } r \to \infty \tag{3.9}$$

for a certain $\gamma < 0$.

Let us consider a compact set of the form $K = K_1 \times [c, \infty)$, $0 \notin K_1$, $c > 0$, and estimate the probability $\mathbf{P}\{\text{hypo}\xi \cap tK \neq \emptyset\}$ as $t \to \infty$. This probability is equal to

$$\begin{aligned}
&\mathbf{P}\{\text{hypo}\xi \cap tK \neq \emptyset\} = \\
&= \mathbf{P}\left\{\sup_{u \in K_1 t} \eta_2 g(u - \eta_1) \geq ct\right\} \\
&= \int_{\mathbb{R}^d} f_1(y_1) dy_1 \int_0^\infty f_2(y_2) \mathbb{I}_{y_2 \sup_{u\in K_1 t} g(u - y_1) \geq ct} dy_2 \\
&= t^{d+1} \int_{\mathbb{R}^d} f_1(ty_1) dy_1 \int_0^\infty f_2(ty_2) \mathbb{I}_{y_2 \kappa(t, y_1) \geq c} dy_2 \\
&= t^{d+1} \int_{\mathbb{R}^d} f_1(ty_1) dy_1 \int_{c/\kappa(t,y_1)}^\infty f_2(ty_2) dy_2 \\
&= t^{d+1} \int_{\mathbb{R}^d} f_1(ty_1) dy_1 \int_1^\infty f_2\left(\frac{tcy_2}{\kappa(t, y_1)}\right) c\kappa(t, y_1)^{-1} dy_2,
\end{aligned}$$

where

$$\kappa(t, y_1) = \sup_{v \in K_1} g((v - y_1)t).$$

If $y \in K_1$, then $\kappa(t, y_1) = g_0$ and

$$\begin{aligned} \mathbf{P}\{\mathrm{hypo}\xi \cap tK \neq \emptyset\} &\geq \frac{c}{g_0} t^{d+1} \int_{K_1} f_1(ty_1) dy_1 \int_1^\infty f_2\left(t \frac{c}{g_0} y_2\right) dy_2 \qquad (3.10) \\ &= I_1. \end{aligned}$$

It follows from properties of regularly varying functions (see Section 1.6) that as $t \to \infty$

$$\begin{aligned} I_1 &\sim \frac{c}{g_0} t^{d+1} t^{\alpha_1 - d} L_1(te_1) \int_{K_1} \phi_1(y_1) dy_1 \left(t \frac{c}{g_0}\right)^{\alpha_2 - 1} L_2(te_2) \int_1^\infty y_2^{\alpha_2 - 1} dy_2 \\ &= -\frac{1}{\alpha_2} t^{\alpha_1 + \alpha_2} L_1(te_1) L_2(te_2) \left(\frac{c}{g_0}\right)^{\alpha_2} \phi_1(y_1) dy_1 \\ &= Q(t) \left(-\frac{1}{\alpha_2}\right) \left(\frac{c}{g_0}\right)^{\alpha_2} \int_{K_1} \phi_1(y_1) dy_1 \qquad (3.11) \end{aligned}$$

for certain $e_1 \in \mathbb{R}^d \setminus \{0\}$ and $e_2 \in (0, \infty)$.

Let us estimate the capacity functional of hypoξ from above. Since $0 \notin K$, the set K^ε misses the origin for sufficiently small ε. Then

$$\mathbf{P}\{\mathrm{hypo}\xi \cap tK \neq \emptyset\} \leq I_2 + I_3,$$

where

$$\begin{aligned} I_2 &= \frac{c}{g_0} t^{d+1} \int_{K_1^\varepsilon} f_1(ty_1) dy_1 \int_1^\infty f_2\left(\frac{tcy_2}{g_0}\right) dy_2 \\ &\sim Q(t) \left(-\frac{1}{\alpha_2}\right) \left(\frac{c}{g_0}\right)^{\alpha_2} \int_{K_1^\varepsilon} \phi_1(y_1) dy_1 \qquad (3.12) \end{aligned}$$

and

$$\begin{aligned} I_3 &= t^{d+1} \int_{\mathbb{R}^d \setminus K_1^\varepsilon} f_1(ty_1) dy_1 \int_{c/\kappa(t,y_1)}^\infty f_2(ty_2) dy_2 \\ &\leq t^{d+1} \int_{\mathbb{R}^d \setminus K_1^\varepsilon} f_1(ty_1) dy_1 \int_{c/\tilde{g}(\varepsilon t)}^\infty f_2(ty_2) dy_2 \\ &= t \int_{(\mathbb{R}^d \setminus K_1^\varepsilon)t} f_1(y_1) dy_1 \int_1^\infty f_2\left(\frac{ty_2 c}{\tilde{g}(\varepsilon t)}\right) \frac{c}{\tilde{g}(\varepsilon t)} dy_2 \\ &\leq \frac{ct}{\tilde{g}(\varepsilon t)} \int_1^\infty f_2\left(\frac{ty_2 c}{\tilde{g}(\varepsilon t)}\right) dy_2 \\ &\sim -\frac{1}{\alpha_2} L_2(te_2) \left(\frac{ct}{\tilde{g}(\varepsilon t)}\right) \\ &= Q_1(t) c^{\alpha_2} \left(-\frac{1}{\alpha_2}\right) \end{aligned}$$

for $e_2 > 0$. It follows from (3.9) that

$$Q_1(t) = o(t^{\alpha_2 - \gamma\alpha_2 + \delta}) \text{ as } t \to \infty$$

for each $\delta > 0$. If $\gamma < -\alpha_1/\alpha_2$, then $Q_1(t) = o(Q(t))$. Hence $\mathbf{P}\{\text{hypo}\xi \cap tK \neq \emptyset\}$ is asymptotically less than

$$Q(t)(1+\lambda_t)\int_{K_1^\varepsilon} \phi_1(y_1)dy_1,$$

where $\lambda_t \to 0$ as $t \to \infty$.

Similarly, for K given in (3.7), we get

$$\begin{aligned} \frac{\mathbf{P}\{\text{hypo}\xi \cap tK \neq \emptyset\}}{Q(t)} &\sim g_0^{-\alpha_2}\left(-\frac{1}{\alpha_2}\right)\sum_{i=1}^{m} c_i^{-\alpha_2}\int_{K_i}\phi_1(y_1)dy_1 \text{ as } t \to \infty \\ &= g_0^{-\alpha_2}\Lambda(K), \end{aligned}$$

where $\Lambda = \mathbf{\Phi}_1 \times \mathbf{\Phi}_2$ is the product of measures $\mathbf{\Phi}_1$ and $\mathbf{\Phi}_2$. Here

$$\mathbf{\Phi}_1(K_1) = \int_{K_1}\phi_1(y)dy$$

is a measure on $\mathbb{R}^d$, and

$$\mathbf{\Phi}_2([c,\infty)) = \int_0^\infty (y_2)^{\alpha_2 - 1}dy_2$$

is a measure on $[0,+\infty)$.

If $K = K_1 \times [c,\infty)$, where $0 \in \text{Int}K$ and $c > 0$, then, for sufficiently small r, we get

$$\begin{aligned} \mathbf{P}\{\text{hypo}\xi \cap tK \neq \emptyset\} &\geq \frac{ct}{g_0}\int_{B_{tr}(0)} f_1(y_1)dy_1 \int_1^\infty f_2\left(\frac{tcy_2}{g_0}\right)dy_2 \\ &\sim \left(\frac{c}{g_0}\right)^{\alpha_2}\left(-\frac{1}{\alpha_2}\right)t^{\alpha_2}L_2(te_2) \text{ as } t \to \infty. \end{aligned}$$

Similarly to the proof of Theorem 4.4.1 we can show that the conditions of Theorem 4.5.1 are valid. Finally, put $a_n = (b_n, c_n)$, where $b_n = kc_n$, $0 < k < \infty$, and

$$c_n = \sup\left\{y\colon\ k^{\alpha_1}y^{\alpha_1+\alpha_2}L_1(te_1)L_2(te_2) \geq 1/n\right\}. \tag{3.13}$$

Then $a_n^{-1} \circ Z_n = a_n^{-1} \circ \text{hypo}\zeta_n$ converges weakly to the U-stable random set Z with the capacity functional given by

$$\tilde{T}(K) = \begin{cases} 1 - \exp\left\{-g_0^{-\alpha_2}\Lambda(K)\right\} &, \quad K \cap (F_1 \cup F_2) = \emptyset \\ 1 &, \quad \text{otherwise} \end{cases}, \tag{3.14}$$

where $F_1 = \{0\} \times [0,\infty)$, $F_2 = \mathbb{R}^d \times \{0\}$ and the measure Λ is defined by

$$\Lambda(K) = \int_K \phi_1(y_1)y_2^{\alpha_2-1}dy_1dy_2,\ K \subset \mathbb{R}^d \times [0,\infty). \tag{3.15}$$

Thus, the random process

$$\zeta_n(u) = c_n^{-1} \max(\xi_1(b_n u), \ldots, \xi_n(b_n u))$$

hypo-converges to the random process ζ such that hypo$\zeta = Z$. The random set Z is constructed in the following way. Take points of the Poisson point process in $\mathbb{R}^d \times \mathbb{R}_+$ with the intensity measure Λ, and attach to each point vertical half-line unbounded from below. Then Z is the union of these half-lines, cf. Norberg (1987), Lyashenko (1986).

Now outline the principal result for another normalization scheme. Put $b_n = 1$ in (3.3), that is

$$\zeta_n(u) = c_n^{-1} \max(\xi_1(u), \ldots, \xi_n(u)).$$

Consider ξ defined in (3.8) and suppose that (3.9) is valid. Now η_1 is not assumed to have regularly varying density. Put

$$c_n = \sup\{y \geq 0 \colon y^{\alpha_2} L_2(te_2) \geq 1/n\}, \; n \geq 1.$$

Then ζ_n hypo-converges to ζ, such that hypoζ has the capacity functional $\tilde{T}$ given by

$$\tilde{T}(K) = \begin{cases} 1 - \exp\{-g_0^{-\alpha_2} \Lambda(K)\} & , \; K \cap F_2 = \emptyset \\ 1 & , \; \text{otherwise} \end{cases}, \tag{3.16}$$

where

$$\Lambda(K) = \int_K f_1(y_1) y_2^{\alpha_2 - 1} dy_1 dy_2, \; K \subset \mathbb{R}^d \times [0, \infty).$$

Note that the capacity functionals (3.14) and (3.16) do not depend on the shape of g, provided (3.9) is valid with $\gamma < -\alpha_1/\alpha_2$. Let us modify (3.8) a little to ensure the dependence on the shape of g. Put

$$\xi(u) = \eta_2 g(u\eta_2^{-1} - \eta_1), \; u \in \mathbb{R}^d, \tag{3.17}$$

where (η_1, η_2) is a random vector in $\mathbb{R}^d \times [0, \infty)$ with positive regularly varying density f (on the product-space $\mathbb{R}^d \times [0, \infty)$) such that $f = \phi L$, for homogeneous ϕ and slowly varying L, and ind$f = \alpha - d - 1$, $\alpha < 0$. Then

$$\text{hypo}\xi = \eta_2 G + \eta_1,$$

where $G =$ hypo$g$ is a non-random set. The function g is supposed to be bounded and have bounded support. Define the multivalued function $M \colon \mathbb{R}^d \times [0, \infty) \longrightarrow \mathcal{K}$ as

$$M(u_1, \ldots, u_{d+1}) = u_{d+1} G + (u_1, \ldots, u_d).$$

Then M is homogeneous. From Theorem 4.4.5 we obtain that the random function

$$\zeta_n(u) = a_n^{-1} \max(\xi_1(a_n u), \ldots, \xi_n(a_n u))$$

(cf. (3.3)) hypo-converges to the random function ζ, such that hypoζ has the capacity functional

$$\tilde{T}(K) = \begin{cases} 1 - \exp\left\{\int_{\mathcal{L}_K} \phi(u)du\right\} & , \quad 0 \notin K \\ 1 & , \quad \text{otherwise} \end{cases},$$

where

$$\mathcal{L}_K = \left\{u \in \mathbb{R}^{d+1}:\ u_{d+1}G + (u_1, \ldots, u_d) \cap K \neq \emptyset\right\}.$$

For example, if $K = \{(x, y)\}$, $x \in \mathbb{R}^d$ and $y > 0$, then

$$\int_{\mathcal{L}_K} \phi(u)du = \int_0^\infty dv \int_{-vG(y/v)+x} \phi(w, v)dw,$$

where $G(a) = \{w \in \mathbb{R}^d: g(w) \geq a\}$.

8.4 Convergence of Random Processes Generated by Approximations of Convex Compact Sets.

Let F be a convex compact set in $\mathbb{R}^d$ with the smooth (twice differentiable) boundary ∂F, and let $\bar{\mathrm{n}}(u)$ be the unit outer normal vector at the point u from ∂F. Furthermore, let $\mathbf{P}$ be a probability measure on F with continuous density f. Consider iid random points $\xi_1, \ldots, \xi_n$, distributed according to the probability law $\mathbf{P}$. Suppose that f is non-vanishing on Int$F$. Then convex hulls of random points $\Xi_n = \text{conv}(\xi_1, \ldots, \xi_n)$ (random polyhedrons) approximate F as n increases.

Various problems of approximation of compact sets by convex hulls of sample points were considered in Schneider (1988), McClure and Vitale (1975), Groeneboom (1988). The main subjects to study were limit theorems for the volume of Ξ_n, distances between F and its polygonal approximation Ξ_n, the number of vertices and other geometric functionals of Ξ_n. Here we prove a limit theorem for the difference between the support functions of Ξ_n and F.

Define $s_{\Xi_n}(u)$ and $s_F(u)$ to be the support functions of Ξ and F respectively, $u \in \mathbb{S}^{d-1}$, and let

$$\begin{aligned} \eta_n(u) &= s_F(u) - s_{\Xi_n}(u) \\ &= \min\left\{s_F(u) - (u \cdot \xi_i):\ 1 \leq i \leq n\right\},\ u \in \mathbb{S}^{d-1}. \end{aligned} \tag{4.1}$$

In this section we prove a limit theorem for the normalized function $a_n^{-1}\eta_n(u)$. Since this function is the minimum of n iid random functions, the epigraph of η_n is the union of epigraphs of functions $s_F(u) - (u \cdot \xi_i)$, $1 \leq i \leq n$. Here $(u \cdot \xi_i)$ is the scalar multiplication in $\mathbb{R}^d$.

Recall that the *epigraph* of the function g is defined as

$$\text{epi}g = \{(u, x):\ g(u) \leq x\},$$

cf. the definition of the hypograph in Section 8.3 Hence epi$\eta_n \subset \mathbb{S}^{d-1} \times \mathbb{R}^1$, and

$$H_n = \text{epi}\eta_n = \bigcup_{i=1}^n A_i,$$

where $A_1, \ldots, A_n$ are iid random sets,

$$A_i = \left\{(u,x) \colon u \in \mathbb{S}^{d-1}, (u \cdot \xi_i) \geq s_F(u) - x\right\}.$$

Consider the sequence of scaled random sets

$$a_n^{-1} \circ H_n = \left\{(u, a_n^{-1}x) \colon (u,x) \in H_n\right\},$$

where $a_n \to 0$ as $n \to \infty$. In fact, the set $a_n^{-1} \circ H_n$ is the epigraph of the random function $a_n^{-1}\eta_n(u)$, $u \in \mathbb{S}^{d-1}$. Similarly to Section 8.3, $a_n^{-1}\eta_n$ is said to *epi-converge* if corresponding epigraphs converge weakly as random closed sets, see also Salinetti and Wets (1986). The weak convergence of $a_n^{-1} \circ H_n$ follows from the pointwise convergence of the corresponding capacity functionals on all compacts, which can be represented as

$$\tilde{K} = \bigcup_{i=1}^{m} K_i \times [0, x_i]. \tag{4.2}$$

Here $K_1, \ldots, K_m$ are compact subsets of $\mathbb{S}^{d-1}$, $x_1, \ldots, x_m > 0$ and $m \geq 1$. We can safely suppose that $K_1, \ldots, K_m$ are canonically closed (with respect to the induced topology on the unit sphere $\mathbb{S}^{d-1}$) and have disjoint interiors.

First, put $\tilde{K} = K \times [0,x]$ and estimate the capacity functional $\mathbf{P}\left\{A_1 \cap t \circ \tilde{K} \neq \emptyset\right\}$ as $t \to 0$. Here K is a canonically closed subset of $\mathbb{S}^{d-1}$, and $t \circ \tilde{K} = K \times [0, tx]$. Introduce also the sets

$$M(K,x) = \{v \in F \colon \inf\{s_F(u) - (u \cdot v) \colon u \in K\} \leq x\}$$

and

$$\begin{aligned} N(K) &= \{v \in \partial F \colon s_F(u) = (v \cdot u) \text{ for a certain } u \in K\} \\ &= \{v \in \partial F \colon \bar{\mathrm{n}}(v) \in K\}. \end{aligned}$$

Then

$$\mathbf{P}\left\{A_1 \cap t \circ \tilde{K} \neq \emptyset\right\} = \mathbf{P}\left(M(K,xt)\right) = \int_{M(K,xt)} f(u)du$$

Furthermore, introduce a set $N(K,x)$ by

$$N(K,x) = \{v - y\bar{\mathrm{n}}(v) \colon v \in N(K), 0 \leq y \leq x\}.$$

Then, for all $\varepsilon > 0$ and sufficiently small t,

$$N(K,xt) \subseteq M(K,xt) \subseteq M(K^\varepsilon, xt), \tag{4.3}$$

where $K^\varepsilon = \{v \in \mathbb{S}^{d-1} \colon B_\varepsilon(v) \cap K \neq \emptyset\}$.

It can be shown that for a certain constant C and every sufficiently small positive t

$$\left|\int_{N(K,xt)} f(u)du - \int_{N(K)} \mu_{d-1}(dv) \int_0^{xt} f(v - y\bar{\mathrm{n}}(v))dy\right| \leq C \sup_{u \in N(K,xt)} f(u)(xt)^2, \tag{4.4}$$

where μ_{d-1} is the $(d-1)$-dimensional Lebesgue measure on ∂F.

Furthermore,

$$\int_{N(K)} \mu_{d-1}(dv) \int_0^{xt} f(v - y\bar{\mathbf{n}}(v))dy = \tag{4.5}$$

$$= \int_{N(K)} \mu_{d-1}(dv) \int_0^{xt} \left[f(v) - (f'(v) \cdot \bar{\mathbf{n}}(v))y + \kappa(v, xt)y^2\right] dy$$

$$= xt \int_{N(K)} f(v)\mu_{d-1}(dv) - \frac{(xt)^2}{2} \int_{N(K)} (f'(v) \cdot \bar{\mathbf{n}}(v))\mu_{d-1}(dv) + o(t^2)$$

as $t \to 0$, where

$$f'(v) = \left(\frac{\partial f}{\partial v_1}, \dots, \frac{\partial f}{\partial v_d}\right).$$

Suppose that $f(v) \neq 0$ for at least one point v from ∂F. Then

$$\mu_{d-1}(\{v \in \partial F \colon f(v) \neq 0\}) > 0.$$

From (4.4), (4.5) we get

$$t^{-1} \int_{N(K,xt)} f(u)du \to x \int_{N(K)} f(v)\mu_{d-1}(dv) \text{ as } t \to 0.$$

It follows from (4.3) that

$$x \int_{N(K)} f(v)\mu_{d-1}(dv) \leq \lim_{t \to 0} t^{-1} \int_{M(K,xt)} f(u)du \leq x \int_{N(K^\epsilon)} f(v)\mu_{d-1}(dv).$$

Similar inequality is valid for K given in (4.2):

$$\sum_{i=1}^m x_i \int_{N(K_i)} f(v)\mu_{d-1}(dv) \leq \lim_{t \to 0} t^{-1}\mathbf{P}\left\{A \cap t \circ \check{K} \neq \emptyset\right\}$$

$$\leq \sum_{i=1}^m x_i \int_{N(K_i^\epsilon)} f(v)\mu_{d-1}(dv).$$

Thus, for $a_n = n^{-1}$, the random closed set $a_n^{-1} \circ H_n$ converges weakly to the random set H having the capacity functional

$$\tilde{T}(K) = 1 - \exp\left\{-\int_{S(K)} f(v)dvdy\right\}, \quad K \subset \mathbb{S}^{d-1} \times [0, +\infty), \tag{4.6}$$

where

$$S(K) = \bigcup_{(u,y) \in K} \{(v, z) \colon v \in N(\{u\}), 0 \leq z \leq y\} \subseteq (\partial F) \times [0, \infty). \tag{4.7}$$

Respectively, $n\eta_n$ epi-converges to the lower semi-continuous random process η such that $\text{epi}\eta = H$.

Consider now an important particular case. Let $\xi_1, \dots, \xi_n$ have the uniform distribution on F. Then the limiting capacity functiona is given by

$$\tilde{T}(K) = 1 - \exp\{-\mu(S(K))/\mu_d(F)\},$$

where $\mu = \mu_{d-1} \times \mu_1$ and μ_1 is the Lebesgue measure on the line.

Now suppose that $f(v) = 0$ for all v from ∂F, and f is continuously differentiable in a certain neighborhood of ∂F. Then

$$\sup_{u \in N(K,xt)} f(u) \to 0 \text{ as } t \to 0,$$

Hence

$$t^{-2} \int_{N(K,xt)} f(u) du \to -\frac{x^2}{2} \int_{N(K)} (f'(v) \cdot \bar{n}(v)) \mu_{d-1}(dv) \text{ as } t \to 0.$$

Thus, for $a_n = n^{-1/2}$, the limiting random set $H = \operatorname{epi}\eta$ (the epi-limit of the random function $a_n^{-1}\eta_n$) has the capacity functional $\tilde{T}$ given by

$$\tilde{T}(K) = 1 - \exp\left\{ \int_{\mathcal{S}(K)} y(f'(v) \cdot \bar{n}(v)) dv dy \right\}, \; K \subset (\mathbb{S}^{d-1} \times [0, +\infty)) \tag{4.8}$$

for $\mathcal{S}(K)$ given by (4.7).

8.5 A Limit Theorem for Intersections of Random Half-Spaces.

For any point $u \in \mathbb{R}^d$ define the corresponding half-space $H(u)$ as

$$H(u) = \left\{ x \in \mathbb{R}^d \colon (x \cdot u) \leq \|u\|^2 \right\}.$$

In this section we consider normalized unions of intersections of random half-spaces defined as

$$X_n = a_n^{-1} \left(H(\xi_1) \cap \cdots \cap H(\xi_n) \right). \tag{5.1}$$

Here $a_n \to 0$ as $n \to \infty$, and $\xi_1, \ldots, \xi_n$ are iid random copies of a certain random vector ξ. Suppose that its density f is regularly varying at zero with index $\alpha - d$ for $\alpha > 0$. It means that the function $\tilde{f}(u) = f(u\|u\|^{-2})$ is regularly varying in the usual sense (see Section 1.6), and $\operatorname{ind}\tilde{f} = d - \alpha$.

Let Y_n be the closure of $\mathbb{R}^d \setminus X_n$. Then $\mathbf{P}\{Y_n \cap K = \emptyset\} = \mathbf{P}\{K \subseteq X_n\}$. Thus, X_n converges weakly to a certain random convex closed set X if and only if Y_n converges weakly to the closure of $\mathbb{R}^d \setminus X$. Moreover, for any compact K

$$\mathbf{P}\{Y \cap K = \emptyset\} = \mathbf{P}\{K \subseteq X\}. \tag{5.2}$$

The random closed set Y_n is the union of complement half-spaces, that is

$$Y_n = a_n^{-1} \left(M(\xi_1) \cup \cdots \cup M(\xi_n) \right),$$

where

$$M(u) = \left\{ x \in \mathbb{R}^d \colon (x \cdot u) \geq \|u\|^2 \right\}. \; u \in \mathbb{R}^d.$$

Let us verify the conditions of Theorem 4.1.3 for the random set $A = M(\xi)$. First, estimate its capacity functional. Evidently,

$$\begin{aligned}\{u\colon K\cap M(u)\neq\emptyset\} &= \{u\in\mathbb{R}^d\colon s_K(u\|u\|^{-1})\geq\|u\|\}\\ &= \{yv\colon v\in S(K), y\geq 0, s_K(v)\geq y\},\end{aligned}$$

where s_K is the support function of K, and

$$S(K)=\{v\in\mathbb{S}^{d-1}\colon s_K(v)\geq 0\}. \tag{5.3}$$

Then

$$\begin{aligned}\mathbf{P}\{A\cap tK\neq\emptyset\} &= \mathbf{P}\{M(\xi)\cap tK\neq\emptyset\}\\ &= \int_{S(K)}dv\int_0^{s_{tK}(v)}y^{d-1}f(yv)dy\\ &= t^d\int_{S(K)}dv\int_0^{s_K(v)}y_1^{d-1}f(y_1vt)dy_1\\ &\sim t^\alpha\frac{1}{\alpha}L(te_0)\int_{S(K)}(s_K(v))^\alpha\phi(v)dv\ \text{ as } t\to 0,\end{aligned}$$

where $e_0>0$, L is slowly varying at zero and ϕ is a homogeneous function, such that $f=\phi L$. Note that similar statements as in Section 1.6 are valid for regularly varying at zero functions.

Put

$$a_n=\inf\left\{t\geq 0\colon t^\alpha L(te_0)\geq\frac{1}{n}\right\}.$$

Then the random set Y_n admits the weak limit Y with the capacity functional $\tilde{T}$ given by

$$\tilde{T}(K)=1-\exp\left\{-\alpha^{-1}\int_{S(K)}(s_K(v))^\alpha\phi(v)dv\right\}. \tag{5.4}$$

From (5.2) we get

$$\mathbf{P}\{K\subseteq X\}=\exp\left\{-\alpha^{-1}\int_{S(K)}(s_K(v))^\alpha\phi(v)dv\right\}. \tag{5.5}$$

Consider two particular cases.

1. If $K=B_r(0)$, then

$$\mathbf{P}\{B_r(0)\subseteq X\}=\exp\left\{-r^\alpha\alpha^{-1}\int_{\mathbb{S}^{d-1}}\phi(v)dv\right\}. \tag{5.6}$$

2. If $K=\{x\}$ is a singleton, then

$$\mathbf{P}\{x\in X\}=\exp\left\{-\alpha^{-1}\int_{S_x^+}\phi(v)(x\cdot v)^\alpha dv\right\}, \tag{5.7}$$

where $S_x^+=\{v\in\mathbb{S}^{d-1}\colon (x\cdot v)\geq 0\}$.

For example, if $d = 2$ and $\phi(v) = C$ for all v from $\mathbb{S}^{d-1}$ (i.e. the function ϕ is circular symmetric) then (5.6) and (5.7) turn into

$$\mathbf{P}\{B_r(0) \subseteq X\} = \exp\{-2\pi C r^\alpha/\alpha\}$$

and

$$\mathbf{P}\{x \in X\} = \exp\left\{-\frac{C}{\alpha}\pi^{1/2}\|x\|^\alpha \frac{\Gamma\left(\frac{\alpha+1}{2}\right)}{\Gamma\left(\frac{\alpha}{2}+1\right)}\right\}.$$

The expected volume of the limiting random set X is finite. It can be evaluated as

$$\begin{aligned} \mathbf{E}\mu(X) &= \int_{\mathbb{R}^d} \mathbf{P}\{x \in X\}\, dx \\ &= \int_{\mathbb{S}^{d-1}} dw \int_0^\infty y^{d-1} \exp\left\{-\alpha^{-1} y^\alpha \int_{S_w^+} \phi(v)(w\cdot v)^\alpha dv\right\} dy \\ &= \Gamma\left(\frac{d}{\alpha}\right)\alpha^{-1}\int_{\mathbb{S}^{d-1}}\left[\alpha^{-1}\int_{S_w^+}\phi(v)(w\cdot v)^\alpha dv\right]^{-d/\alpha} dw. \end{aligned}$$

Intersections of random half-spaces appear in linear programming problems with random constraints. Consider n random constraints given by

$$\eta_{i1}x_1 + \cdots + \eta_{id}x_d \le \eta_{i(d+1)}, \quad 1 \le i \le n, \tag{5.8}$$

where $(\eta_{i1}, \ldots, \eta_{id}, \eta_{i(d+1)})$, $1 \le i \le n$, are iid random vectors in $\mathbb{R}^{d+1}$. Put

$$\xi_i = \frac{(\eta_{i1}, \ldots, \eta_{id})}{(\eta_{i(d+1)})^{1/2}}.$$

Then X_n, given by (5.1), coincides with the normalized set of all admissible solutions of (5.8). If the density of the random vector ξ is regularly varying at zero, $\operatorname{ind} f > -d$, then the weak limit X of the set of solutions of (5.8) exists and its distribution is given by (5.5).

It is particularly interesting to obtain limit theorems for maxima values of linear functionals on X_n. Let

$$(h\cdot x) = h_1x_1 + \cdots + h_dx_d$$

be a linear functional on $\mathbb{R}^d$, where $h = (h_1, \ldots, h_d)$. For simplicity suppose that $\|h\| = 1$. Then

$$\sup_{x\in X}(h\cdot x) = s_X(h),$$

and $s_{X_n}(h)$ converges weakly to $s_X(h)$. The exact distribution of $s_X(h)$ is very difficult to evaluate, since, in fact, we do not know the capacity functional of X, but only probabilities given by (5.5).

Let us estimate the distribution function of $s_X(h)$ from below in the following way

$$\mathbf{P}\{s_X(h) \ge a\} \ge \mathbf{P}\{ha \in X\}.$$

It follows from (5.7) that

$$\mathbf{P}\{s_X(h) \ge a\} \ge \exp\left\{-\alpha^{-1}a^\alpha \int_{S_h^+}\phi(v)(h\cdot v)^\alpha dv\right\}.$$

Then the expectation of $s_X(h)$ is estimated from below as follows

$$\begin{aligned}\mathbf{E}s_X(h) &\geq \int_0^\infty \exp\left\{-\alpha^{-1}y^\alpha \int_{S_h^+} \phi(v)(h\cdot v)^\alpha dv\right\} dy \\ &= \alpha^{-1}\Gamma\left(\frac{1}{\alpha}\right)\left[\alpha^{-1}\int_{S_h^+}\phi(v)(h\cdot v)^\alpha dv\right]^{-1/\alpha}\end{aligned}$$

References

AMBARTZUMIAN, R.V. (1990) *Factorization Calculus and Geometric Probability.* Cambridge Univ. Press, Cambridge.

ARAUJO, A. AND E.GINE (1980) *The Central Limit Theorem for Real and Banach Valued Random Variables.* Wiley, New York.

ARROW, K.J AND F.H.HAHN (1971) *General Competitive Analysis.* San Francisco, Holden-Day.

ARTSTEIN, Z. (1984) Limit laws for multifunctions applied to an optimization problem. *Lect. Notes Math.*, **1091**, 66-79.

ARTSTEIN, Z. AND R.A.VITALE (1975) A strong law of large numbers for random compact sets. *Ann. Probab.*, **3**, 879-882.

ATTOUCH, H. AND R.J.-B.WETS (1990) Epigraphical processes: law of large numbers for random LSC functions. *Sem. Anal. Convexe.*, 20, Exp. No 13, 29 p.

AUBIN, J.-P. AND I.EKELAND (1984) *Applied Nonlinear Analysis.* Wiley, New York etc.

AUBIN, J.-P. AND H.FRANKOWSKA (1990) *Set-Valued Analysis.* Birkhauser, Boston.

AUMANN, R. (1965) Integrals of set-valued functions. *J. Math. Anal. Appl.*, **12**, 1-12.

BADDELEY, A. (1991) Hausdorff metric for capacities. Unpublished.

BALKEMA, A.A. AND S.I.RESNICK (1977) Max-infinite divisibility. J. Appl. Probab., **14**, 309-319.

BERG, C., CHRISTENSEN, J.P.R. AND P.RESSEL (1984) *Harmonic Analysis on Semigroups.* Springer, Berlin etc.

BHATTACHARIA,R.N. AND R.RANGA RAO (1976) *Normal Approximation and Asymptotic Expansions.* Wiley, New York.

BILLINGSLEY, P. (1968) *Convergence of Probability Measures.* Wiley, New York.

CHOQUET, G. (1953/54) Theory of capacities. *Ann. Inst. Fourier*, **5**, 131-295.

CLARKE, H. (1983) *Optimization and nonsmooth analysis.* Wiley, New York.

CRESSIE, N. (1979) A central limit theorem for random sets. *Z. Wahrscheinlichkeitstheorie.*, **49**, 37-47.

CRESSIE, N. AND G.M.LASLETT (1987) Random set theory and problems of modeling. *SIAM Review*, **29**, 557-574.

DAVIS, R.A., MULROW, E. AND S.I.RESNICK (1987) The convex hull of a random sample in $\mathbb{R}^2$. *Comm. Stat. Stochastic Models*, **3**(1), 1-27.

DAVIS, R.A., MULROW, E. AND S.I.RESNICK (1988) Almost sure limit sets of random samples in $\mathbb{R}^d$. *Adv. Appl. Probab.*, **20**, 573-599.

DUDLEY, R.M. (1984) A course on empirical processes. *Lect. Notes Math.*, **1097**, 1-142.

FELLER, W. (1971) *An Introduction to Probability Theory and Its Applications.* **2**, 2nd edn. Wiley, New York.

GALAMBOS, J. (1978) *The Asymptotic Theory of Extreme Order Statistics.* Wiley, New York.

GERRITSE, B. (1993) Integration with respect to capacities and applications in large-deviation theory. In: *22nd Conf. Stochastic Proc. Their Appl.*, Amsterdam, 21-25 June, 1993. Abstracts, p.55.

GERRITSE, G. (1986) Supremum self-decomposable random vectors. *Probab. Th. Rel. Fields*, **72**, 17-33.

GERRITSE, G. (1990) *Self-decomposable distributions in continuous lattices.* Catholic University, Nijmegen. Report n.9028.

GINE, E. AND M.G.HAHN (1985a) The Levy-Hincin representation for random compact convex subsets which are infinitely divisible under Minkowski addition. *Z. Wahrscheinlichkeitstheorie.*, **70**, 271-287.

GINE, E. AND M.G.HAHN (1985b) Characterization and domains of attraction of p-stable compact sets. *Ann. Probab.*, **13**, 447-468.

GINE, E., HAHN, M.G. AND J.ZINN (1983) Limit theorems for random sets: application of probability in Banach space results. *Lect. Notes Math.*, **990**, 112-135.

GINE, E., HAHN, M.G. AND P.VATAN (1990) Max-infinitely divisible and max-stable sample continuous processes. *Probab. Th. Rel. Fields*, **87**, 139-165.

GROENEBOOM, P. (1988) Limit theorems for convex hulls. *Probab. Th. Rel. Fields.* **79**, 327-368.

HAAN, L. DE (1970) *On Regular Variation and Its Application to the Weak Convergence of Sample Extremes.* Math. Centre Tracts **32**, Mathematics Centre, Amsterdam.

HAAN, L. DE AND E.OMEY (1983) Integrals and derivatives of regularly varying functions in R^d and domains of attraction of stable distributions. II. *Stoch. Proc. Appl.*, **16**, 157-170.

HAAN, L. DE AND S.I.RESNICK (1987) Derivatives of regularly varying functions in $\mathbb{R}^d$ and domains of attraction of stable distributions. *Stoch. Proc. Appl.*, **8**, 349-355.

HAAN, L. DE AND S.I.RESNICK (1987) On regular variation of probability densities. *Stoch. Process. Appl.*, **25**, 83-93.

HENGARTNER, W. AND R.THEODORESCU (1973) *Concentration Functions.* Academic Press, New York - London.

HIAI, F. (1984) Strong laws of large numbers for multivalued random variables. *Lect. Notes Math.*, **1091**, 160-172.

HIAI, F. AND H.UMEGAKI (1977) Integrals, conditional expectations, and martingales of multivalued functions. *J. Multivar. Anal.*, **7**, 149-182.

HUBER, P.J. (1981) *Robust Statistics.* Wiley, New York.

ITO, K. AND H.P.MCKEAN (1965) *Diffusion Processes and Their Sample Paths.* Springer-Verlag, Berlin etc.

KALASHNIKOV, V.V. AND S.T.RACHEV (1988) *Mathematical Methods of Constructions of Stochastic Queuing Models.* Nauka, Moscow (in Russian).

KENDALL, D.G. (1974) Foundations of a theory of random sets. In: *Stochastic Geometry* /Eds.:E.F.Harding and D.G.Kendall, Wiley, New York.

KRUSE, R. (1987) On the variance of random sets. *J. Math. Anal. Appl.*, **122**, 469-473.

LANDKOF, N.C. (1966) *Foundations of Modern Potential Theory.* Nauka, Moscow (in Russian).

LEADBETTER, M.R., LINDGREN, G. AND H.ROOTZEN (1986) *Extremes and Related Properties of Random Sequences and Processes.* Springer-Verlag, Berlin.

LYASHENKO, N.N. (1983) Statistics of random compacts in the Euclidean space. *J. Soviet Math.*, **21**, 76-92.

LYASHENKO, N.N. (1983) Weak convergence of step-functions in the space of closed sets. *Zapiski Nauch. Seminarov LOMI*, **130**, 122-129 (in Russian).

LYASHENKO, N.N. (1986) Graphs of random processes as random sets. *Th. Probab. and Its Appl.*, **31**, 81-90 (in Russian).

MATHERON, G.(1975) *Random Sets and Integral Geometry.* Wiley, New York.

MCCLURE, D.E. AND R.A.VITALE (1975) Polygonal approximation of plane convex bodies. *J. Math. Anal. Appl.*, **51**, 326-358.

MOLCHANOV, I.S. (1984) A generalization of the Choquet theorem for random sets with a given class of realization. *Th. Prob. Math. Statist.*, **28**, 99-106.

MOLCHANOV, I.S. (1987) Uniform laws of large numbers for empirical capacity functionals of random closed sets. *Th. Probab. and Its Appl.*, **32**, 556-559.

MOLCHANOV, I.S. (1991) Random sets: survey on some results and applications. *Ukrainian Math. J.*, **43**, 1477-1487.

MOLCHANOV, I.S., (1992) Characterization of union-stable random closed sets. *Th. Prob. Math. Statist.*, **46**, 114-120 (in Ukrainian).

MOLCHANOV, I.S. (1993a) On the convergence of pointwise maxima of special random functions. In: *Stability Problems for Stoch. Models.* Proceedings of the Intern. Conf. held at Perm Russia, June 1-6, 1992. Eds.: V.M.Zolotarev et al., TVP Science Publisher (Ser. Frontiers in Pure and Applied Probability, vol.3).

MOLCHANOV, I.S. (1993b) On distributions of random closed sets and expected convex hulls. *Statistics & Prob. Letters*, **17**.

MOLCHANOV, I.S. (1993c) Strong law of large numbers for unions of random closed sets. *Stoch. Proc. Appl.* (in print).

MOLCHANOV, I.S. (1993d) Limit theorems for convex hulls of random sets. *Adv.*

Appl. Prob. **25**, 395-414.

MOLCHANOV, I.S. (1993e) Limit theorems for unions of random sets with multiplicative norming. *Th. Probab. and Its Appl.* **38**.

NORBERG, T. (1984) Convergence and existence of random set distributions. *Ann. Probab.*, **12**, 726-732.

NORBERG, T. (1985) *Random Sets and Capacities, with Applications to Extreme Value Theory.* Ph.D. Thesis, Department of Mathematics, Gteborg University.

NORBERG, T. (1986a) *On the existence and convergence of probability measures on continuous semi-lattices.* Techn. Report n.148, Center for Stochastic Processes, Univ. of North Carolina.

NORBERG, T. (1986b) Random capacities and their distributions. *Probab. Th. Relat. Fields*, **73**, 281-297.

NORBERG, T. (1987) Semicontinuous processes in multi-dimensional extreme-value theory. *Stoch. Process. Appl.*, **25**, 27-55.

NORBERG, T. (1989) Existence theorems for measures on continuous posets, with applications to random set theory. *Math. Scand.*, **64**, 15-51.

NORBERG, T. AND W.VERVAAT (1989) *Capacities on non-Hausdorff spaces.* Preprint 1989-11 ISSN 0347-2809, Department of Mathematics, Chalmers University of Technology and the University of Goteborg, Goteborg.

PANCHEVA, E. (1985) Limit theorems for extreme order statistics under non-linear normalization. *Lect. Notes Math.*, **1155**, 284-309.

PANCHEVA, E. (1988) Max-stability. *Th. Prob. Its Appl.*, **33**, 167-170.

PAPAGEORGIOU, N.S. (1987) Random functional-differential inclusions with nonconvex right hand side in a Banach space. *Comment. Math. Univ. Carolinae*, **28**, 649-654.

PURI, M.L. AND D.A.RALESCU (1985) Limit theorems for random compact sets in Banach space. *Math. Proc. Camb. Phil. Soc.*, **97**, 151-158.

RACHEV, S.T. (1986) Pevy-Prokhotrov distance in a space of semicontinuous set functions. *J. Soviet Math.*, **34**, 112-118.

RACHEV, S.T. (1991) *Probability Metrics of Stochastic Models.* Wiley, Chichester.

RESNICK, S.I. AND R.TOMKINS (1973) Almost sure stability of maxima. *J. Appl. Probab.*, **10**, 387-401.

RESNICK, S.I. (1986) Point processes, regular variation and weak convergence. *Adv. Appl. Prob.*, **18**, 66-138.

ROBBINS, H.E. (1944) On the measure of a random set. I. *Ann. Math. Statist.*, **15**, 70-74.

ROCKAFELLAR, R.T. AND R.J.-B.WETS (1984) Variational systems, an introduction. *Lect. Notes Math.*, **1091**, 1-54.

SALINETTI, G. (1987) Stochastic optimization and stochastic processes: the epigraphical approach. *Math. Res.*, **35**, 344-354.

SALINETTI, G. AND R.J.-B.WETS (1986) On the convergence in distribution of measurable multifunctions (random sets), normal integrands, stochastic processes and stochastic infima. *Math. Oper. Res.*, **11**, 385-419.

SANTALÓ, L.A. (1976) *Integral Geometry and Geometric Probability.* Addison-Wesley, Reading (Mass.).

SCHNEIDER, R. (1988) Random approximations of convex sets. *J. Microsc.*, **151**, 211-227.

SENETA, E. (1976) *Regularly Varying Functions.* Springer-Verlag, Berlin etc.

SERRA, J.-P. (1982) *Image Analysis and Mathematical Morphology.* Academic Press, London.

STOYAN, D. (1989) On means, medians and variances of random compact sets. *Math. Res.*, **51**, 99-104.

STOYAN, D., KENDALL, W.S. AND J.MECKE (1987) *Stochastic Geometry and Its Applications.* Akademie-Verlag, Berlin.

STOYAN, D. AND H.STOYAN (1992) *Fraktale – Formen – Punktfelder: Methoden der Geometrie-Statistik.* Berlin, Akademie Verlag.

TRADER, D.A. (1981) *Infinitely Divisible Random Sets.* Ph.D. Dissertation. Carnegie-Mellon University.

URYSON, P.S. (1924) Dependence between the mean breadth and the volume of convex bodies in the n-dimensional space. *Matem. Zbornik.*, **31**, 477-486 (in Russian).

VERVAAT, W. (1988) *Random Upper Semicontinuous Functions and Extremal Processes.* Rept.Cent. Math. and Comput. Sci. MS-R88901, Amsterdam.

VITALE, R.A. (1983) Some developments in the theory of random sets. *Bull. Int. Statist. Inst.*, **50**, 863-871.

VITALE, R.A. (1984) On Gaussian random sets. In: *Stochastic Geometry, Geometric Statistics, Stereology* /Eds.: R.V.Ambartzumian, W. Weil; Teubner, 222-224.

VITALE, R.A. (1987) Expected convex hulls, order statistics and Banach space probabilities. *Acta Appl. Math.*, **9**, 97-102.

VITALE, R.A. (1988) An alternate formulation of mean value for random geometric figures. *J. Microsc.*, **151**, 197-204.

VITALE, R.A. (1990) The Brunn-Minkowski inequality for random sets. *J Multivar. Analysis*, **33**, 286-293.

WAGNER, D. (1979) Survey of measurable selection theorem: an update. *Lect. Notes Math.*, **794**, 176-219.

WEIL, W. (1982) An application of the central limit theorem for Banach-space-valued random variables to the theory of random sets. *Z. Wahrscheinlichkeitstheorie.*, **49**, 37-47.

YAKIMIV, A.L. (1981) Multivariate Tauberian theorems and their applications to the branching Bellman-Harris processes. *Matem. Zbornik*, **115**, 453-467 (in Russian).

ZOLOTAREV, V.M. (1979) Ideal metrics in the problems of probability theory. *Austral. J. Statist.*, **21**, 193-208.

ZOLOTAREV, V.M. (1986) *Modern Theory of Summation of Independent Random Variables.* Nauka, Moscow (in Russian).

Index

Author Index

Subject Index

Printing: Weihert-Druck GmbH, Darmstadt
Binding: Buchbinderei Schäffer, Grünstadt